SpringerBriefs in Molecular Science

Electrical and Magnetic Properties of Atoms,
Molecules, and Clusters

Series editor

George Maroulis, Patras, Greece

W0230296

More information about this series at http://www.springer.com/series/11647

Stavros C. Farantos

Nonlinear Hamiltonian Mechanics Applied to Molecular Dynamics

Theory and Computational Methods for Understanding Molecular Spectroscopy and Chemical Reactions

 Springer

Stavros C. Farantos
Department of Chemistry
University of Crete
Iraklion
Greece

and

Institute of Electronic Structure and Laser
Foundation for Research and Technology-Hellas
Iraklion
Greece

ISSN 2191-5407   ISSN 2191-5415   (electronic)
ISBN 978-3-319-09987-3   ISBN 978-3-319-09988-0   (eBook)
DOI 10.1007/978-3-319-09988-0

Library of Congress Control Number: 2014947684

Springer Cham Heidelberg New York Dordrecht London

Printed on acid-free paper

Springer is part of Springer Science+Business Media (www.springer.com)

Preface

The first being so, and so the second,
The third and forth deduced we see;
And if there were no first and second,
Nor third nor fourth would ever be.
from "Faust" by Goethe

The birth of nonlinear mechanics and particularly their geometrical approach is usually attributed to Henri Poincaré[1] and they were fully developed in the second half of the twentieth century. New concepts were introduced and theorems were proved by mathematicians, thus putting the foundations for understanding *ordered* (stable) and *chaotic* (deterministic) motions of nonlinear dynamical systems. Now, a plethora of diverse complex systems are studied by nonlinear mechanics, which play the role of a unifying theory. In this success the rapid development of computers unequivocally played a catalytic role. Molecules are complex nonlinear dynamical systems and several research groups around the world rushed at the same time to investigate the implications of the theory of chaos could have in the dynamics of molecules. Since atoms and molecules are treated by quantum mechanics questions of how to interpret nonlinear classical mechanical behaviours in the quantum world, they quickly became a hot subject in the 1970s giving birth to what was named *quantum chaos*.

Nevertheless, in spite of this flourishing and productive period of nonlinear mechanics, their ideas seem to have no impact on the large community of theoretical and computational chemists. Instead, most of the efforts of computational chemists are still devoted to producing *potential energy surfaces* the stationary points of which, minima, saddles, maxima, as well as minimum energy pathways provide the theoretical background for explaining experimental spectroscopic and

---

[1] Henri Poincaré. Les Méthodes Nouvelles de la Mécanique Céleste, Vols 1–3. Gauthiers-Villars, Paris, 1892, 1893, 1899. (English translation edited by D. Goroff, published by the American Institute of Physics, New York, 1993.)

reaction dynamics observations. In the Born–Oppenheimer approximation,[2] the potential energy surface is used to calculate the forces among nuclei and classical rather than quantum mechanics are the main theories to study the dynamics of molecules with a relatively large number of atoms. Classical mechanics combined with statistical mechanics for extracting average quantities consist what is known today as Molecular Dynamics.

The limited interest of chemists in nonlinear mechanics is understood if one considers that even today to produce a reliable potential energy surface for a medium size polyatomic molecule (up to five atoms) requires substantial effort. There are also some basic reasons for the reluctance of chemists to introduce nonlinear mechanics to their ammunition in investigating molecular dynamics. Hamiltonian mechanics and their geometrical interpretations are essential for nonlinear mechanics, topics that still remain out of the chemists curriculum in postgraduate studies. Needless to say, the lack of an introductory book in nonlinear mechanics for chemists, significantly contributes to their limited interest in this field. The aim of the present book is to partially fill this gap. On the other hand, in the last decades there has been enormous progress in experimental techniques, which provide details at the level of single molecule quantum states. Methods for spectroscopically assigning highly excited vibrational states of reactant and product molecules in chemical reactions have been developed. Molecular beams, lasers, and ion-imaging technologies have contributed to even follow in real time how a chemical bond in a molecule breaks or is formed. For these achievements in reaction dynamics Ahmed Zewail was awarded the Nobel prize in 1999.

The book focuses on the basic definitions, theorems, and computational algorithms developed by nonlinear mechanics with examples from small polyatomic molecules. No mathematical rigor is claimed and by all means this is not another book on nonlinear mechanics. Emphasis is given to numerical methods which can be extended to many degrees of freedom systems, thus, assisting one to apply them to realistic molecular potentials. Most of molecular theories in chemistry consider molecules as *conserved Hamiltonian systems*, hence, we present the theory of nonlinear mechanics pertinent to this class of dynamical systems.

The book is organized along the following directions. After the introduction in Chap. 1, which also gives a historical overview of the field and its current status, Chap. 2 presents a brief introduction to Hamiltonian mechanics. An effort is made to present the theory from the analytical mechanics point of view, which unveils the geometrical characteristics of the theory, such as its symplectic symmetry. In Chap. 3 dynamical systems are introduced and the basic invariant structures, the main subject of nonlinear mechanics, are presented by numerically studying simple one-, two- and three-dimensional model potentials. Chapter 4 deals with quantum and semiclassical molecular mechanics. Algorithms and numerical methods for treating classical nonlinear and associated quantum mechanical equations of motion

---

[2] M. Born and J.R. Oppenheimer. On the Quantum Theory of Molecules. *Ann. Physik*, 84:457, 1927. Translated by S.M. Blinder, http://en.wikipedia.org/wiki/Born-Oppenheimer_approximation.

are discussed in Chap. 5 with emphasis to methods developed by us. Chapter 6 is devoted to applications, which demonstrate how the numerical codes serve to study real polyatomic molecules. Finally, in Chap. 7, some ideas of how progress in computer technology will affect the field of nonlinear molecular dynamics are put forward. An extended Appendix which describes basic mathematical concepts and theorems of modern mathematical analysis of manifolds supplements the book. The aim is not to apply mathematical rigor, but to exempt the reader from the need to look for definitions and explanations of these, admittedly, not very familiar to the chemists concepts. In the book terms referred to definitions, and when they appear for first time, are written with italic fonts, whereas those terms which are also illustrated in the Appendix are written by italic-bold letters.

The book is based on the author's years of research and the work of his Ph.D. students and collaborators around the world, in the field of nonlinear mechanics applied to molecular dynamics. Manolis Founargiotakis, Rita Prosmiti, and Stamatis Stamatiadis completed their Ph.D. theses by developing parts of the software and applying it to several molecules. The postdoctoral fellows Raul Guantes and Jaime Suarez contributed together with Stamatis Stamatiadis to the development of the variable order finite difference codes for discretizing the Schrödinger equation. George Contopoulos and the late Chronis Polymilis, astronomers from university of Athens, assisted to transfer knowledge on periodic orbits from the macroscopic world of galaxies to the microscopic world of molecules by studying two- and three-dimensional model potentials, common to both molecular and galactic dynamics. Reinhard Schinke from Max-Planck Institut für Dynamik and Selbstorganisation in Göttingen has been a collaborator for almost twenty years, steadily provided me with interesting molecules, which showed spectroscopic unidentified fingerprints implying unexpected dynamics. I believe, that most of them found an explanation by treating these molecules as nonlinear dynamical systems. Hua Guo from University of New Mexico has also been a provider of accurate potential energy surfaces and results from accurate quantum dynamics of molecules showing interesting experimental behaviours. Howard Taylor from the University of Southern California introduced me to the technology of classical autocorrelation functions, a powerful method for exploring the molecular phase space. Recently, Vladimir Tyuterev from University of Reims and then his student Frederic Mauguiére are two of the latest collaborators. Projects on investigating the isotopic mass effect in the spectroscopy of highly excited molecules, such as water, by periodic orbits were carried out. Vangelis Daskalakis and Constantine Varotsis from Cyprus University of Technology are the collaborators with whom the work on the active site of the enzyme cytochrome c oxidase by periodic orbits was materialized.

Last, but not least, I am grateful to Stephen Wiggins and Gregory Ezra for their encouragement and useful comments in structuring this brief book. My collaboration with them and Steve's postdoctoral fellows, Peter Collins and Frederic Mauguiére, for more than a year now, has helped me to enlighten many aspects of phase space geometry related to reaction dynamics. Our 'skype meeting', almost one every week, has become for me an exciting scientific event.

The literature in the field of nonlinear mechanics is really vast. Inevitably, the references to articles and books cited in this book are those which had the most influence to the author, or stating it better, with which the author came across. By no means papers and books not mentioned here are of limited significance for the field.

Iraklion, Crete, Greece, July 2014   Stavros C. Farantos

Contents

Chapter 1
Introduction and Overview

The study of dynamical properties of molecules entails a quantum mechanical approach. Considering molecules as aggregates of electrons and nuclei a method to simplify the Schrödinger equation is to separate the motion of the fast moving electrons from that of the heavy nuclei as was proposed by Born and Oppenheimer [2] only a year after the birth of quantum mechanics. They applied the *adiabatic approximation* to separate *fast from slow motions* and to split the molecular Schrödinger equation into two uncoupled equations, one, which treats only the motions of electrons with the nuclei frozen at a specific *geometrical configuration*, and a second one, that treats the nuclei moving under the average electronic potential. Since then, the electronic adiabatic Schrödinger equation is numerically solved by a plethora of robust and versatile in accuracy algorithms to produce what is known as Born–Oppenheimer (adiabatic) eigenstates for molecules with hundreds of atoms. The field of Computational Chemistry that encompasses methods and computer codes for solving the *electronic Schrödinger equation* is named *Quantum Chemistry*. The seminal paper of Born–Oppenheimer could be considered as inaugurating the birth of the field of *Chemical Physics*.

Solving the nuclear equations of motion in specific electronic states consists the second pillar of Computational Chemistry named *Molecular Dynamics*. Although, the goal is always to carry out a quantum molecular dynamical study, the fact that we deal with heavy particles legitimate one to use semiclassical or pure classical mechanical theories for investigating molecular vibrations, rotations and chemical reactions. Molecules with thousand of atoms can be treated by solving Newton's or Hamilton's equations of motion and detailed dynamics, equilibrium and non-equilibrium statistical mechanical properties can be extracted. In spite of the triumph of quantum mechanics, research on semiclassical theory and comparisons of quantum to classical mechanics have never stopped almost a century after the discovery of quantum mechanics.

The separation of electronic from nuclear motion within the adiabatic approximation had a profound impact to chemistry by introducing the concept of *Potential Energy Surface* (PES). This is a function of $F_v = 3N - 6$ internal nuclear coordinates (for example the bond lengths and angles between them called *valence coordinates*),

© The Author(s) 2014

S.C. Farantos, *Nonlinear Hamiltonian Mechanics Applied to Molecular Dynamics*,
SpringerBriefs in Electrical and Magnetic Properties of Atoms, Molecules, and Clusters,
DOI 10.1007/978-3-319-09988-0_1

on the **configuration manifold** of a molecule with N atoms.[1] PES is the potential that governs the motions of nuclei at a particular electronic state of the molecule. Minima on this *hypersurface* correspond to stable isomers, whereas saddles are approximately used to define transition states for the isomerization or dissociation of the molecule. Studying reactions the knowledge of an extended region or even the global PES is required, whereas to assign low excitation vibrational spectra of a molecule the determination of the PES in the region around the minimum of a stable isomer is in most cases enough. Expanding the molecular potential in a Taylor series around the minimum, the first non-zero terms of the polynomial, apart from a constant term, involve the second derivatives of the potential function. Thus, by keeping only the second order terms in the expansion and employing linear transformations to **normal coordinates** we can always write a Hamiltonian as a sum of $3N - 6$ ($3N - 5$ for linear molecules) independent harmonic oscillators, i.e., the PES in normal coordinates is approximated as sum of quadratic terms. Hence, the forces are linear functions of the normal coordinates and similarly the corresponding differential equations that describe the equations of motion. Higher order terms in the Taylor expansion of the PES will result in higher order terms in normal coordinates (cubic, quartic, etc.), and the classical mechanical equations of motion are then named *nonlinear*. Although, we can analytically solve the linear equations of motion in normal coordinates for a multi-dimensional system, this is not any longer feasible for the nonlinear equations. Under special regimes perturbation methods, such as the powerful method of **normal forms**, could separate the coupled differential equations to F_v 1D equations, thus, approximating **non-integrable** systems by **integrable** ones. Nevertheless, in most cases both the classical and quantum equations of motion must be solved numerically.

1.1 Nonlinear Mechanics and Molecular Dynamics

It is remarkable that one of the first applications of MANIAC-I, the first computer constructed in Los Alamos in 1950s, was a project proposed by Fermi, Pasta and Ulam and involved the solution of Newton's equations for a one-dimensional, harmonic chain of particles weakly perturbed by nonlinear forces with the Hamiltonian

$$H = \frac{1}{2} \sum_{k=1}^{N-1} P_k^2 + \frac{1}{2} \sum_{k=0}^{N-1} (Q_{k+1} - Q_k)^2 + \frac{a}{3} \sum_{k=0}^{N-1} (Q_{k+1} - Q_k)^3, \qquad (1.1)$$

and $Q_0 = Q_N = 0$.

As it is now known, the Fermi-Pasta-Ulam-Tsingou (FPUT) model (Fig. 1.1) was used to answer questions of energy equipartition and ergodicity in a many particle system, which are usually surmised in statistical mechanics [10].

---

[1] The dimension of the PES for linear molecules is $3N - 5$.

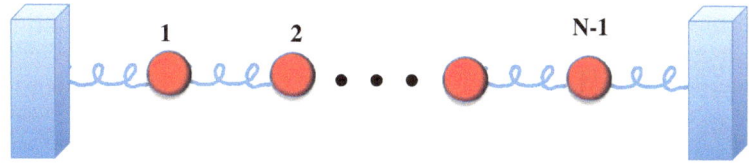

Fig. 1.1 The Fermi-Pasta-Ulam-Tsingou model of N-particles in a harmonic chain and perturbed by cubic nonlinear potential terms (Eq. 1.1)

The investigators carried out calculations for $N = 32$ particle chain and $a = 1/4$ by integrating the classical mechanical equations of motion in Cartesian coordinates. However, the analysis of the results was done in normal coordinates. To the surprise of investigators, plotting the energy of the different normal modes in time they saw recurrences which revealed *quasiperiodic* behaviour and not energy randomization among all degrees of freedom as it was expected for *chaotic* trajectories. Specifically, they found that the energy in the cubic nonlinear terms never exceeded about 10% of the total energy.

For quasiperiodic trajectories in the time evolution of the system more quantities than solely the total energy are conserved. If there are as many constants of motion as the number of degrees of freedom, then the system is integrable. In this case, transformations of the normal mode coordinates to cyclic variables exist, named **action-angle** variables. Non-integrable systems have chaotic trajectories and conserve only global constants of motion, such as the total energy, and the assumption of ergodicity may be adopted.

For polyatomic molecules the PES is a nonlinear, multivariable function. Even assuming that we have an analytical function for the PES in internal coordinates, which is often true for triatomic and tetratomic molecules, to find the kinetic part of the Hamiltonian involves cumbersome calculations to extract the conjugate momenta or the corresponding differential operators in the quantum case. Therefore, it is not surprising that in most of the computer codes available for solving the equations of motion, either in classical or in quantum mechanics, we prefer to employ Cartesian coordinate systems. On the other hand, adopting curvilinear coordinates, such as valence or Jacobi, in molecules and by doing the proper transformations, which lead to separable or near separable new variables, is a vital step to predict and elucidate quasiperiodic (regular) behaviour of the system in all or some degrees of freedom. As a mater of fact, this is the strategy which is followed when we deal with harmonic potentials. Let us see how this works in the case of carbon dioxide by considering for simplicity only the two stretching vibrational modes of this linear molecule, the antisymmetric and symmetric stretches.

Taking the molecule to lie along the z-axis the displacements from the equilibrium positions of the three atoms are denoted as $(\delta Z_{O_1}, \delta Z_C, \delta Z_{O_2})$. The transformation from Cartesian to normal mode coordinates are described in many textbooks [13]. Here, we only sketch out the solution. Although, the equations of motion in normal coordinates are decoupled, a better understanding of the physical motions of the molecule is obtained by further transforming to what we call action-angle variables, to be defined properly in Chap. 4. Furthermore, we shall see that by transforming to action-angle variables we can semiclassically quantize the vibrational modes of the molecule. Hence, considering the existence of successive inverse transformations from action-angle, $(I_1, \phi_1, I_2, \phi_2)$, to normal mode, (Q_{ss}, Q_{as}) to Cartesian coordinates, we can write symbolically

$$\begin{aligned}
\delta Z_{O_1} &= g_{Z_{O_1}}[Q_{ss}(\phi_1; I_1), Q_{as}(\phi_2; I_2)] \\
\delta Z_C &= g_{Z_C}[Q_{ss}(\phi_1; I_1), Q_{as}(\phi_2; I_2)] \\
\delta Z_{O_2} &= g_{Z_{O_2}}[Q_{ss}(\phi_1; I_1), Q_{as}(\phi_2; I_2)]
\end{aligned} \tag{1.2}$$

The importance of these transformations is the realization of the *geometry of the motion in phase space*. In Fig. 1.2 we show that in action-angle variables, quasiperiodic trajectories integrated in time lie on 2D surfaces embedded in 4D phase space, with geometry that of a ***torus***. A torus is the the product of two circles, $T^2 = S^1 \times S^1$ (yellow lines). The regular motions of the decoupled system of harmonic oscillators survive even when the nonlinear coupling terms in the potential are switched on. This is proved by the renowned theorem of *Kolmogorov-Arnold-Moser (KAM)* [30], which states that most of the tori will remain, although slightly deformed, and only at large couplings or high excitation energies the tori will be destroyed and chaos will occupy most of the phase space.

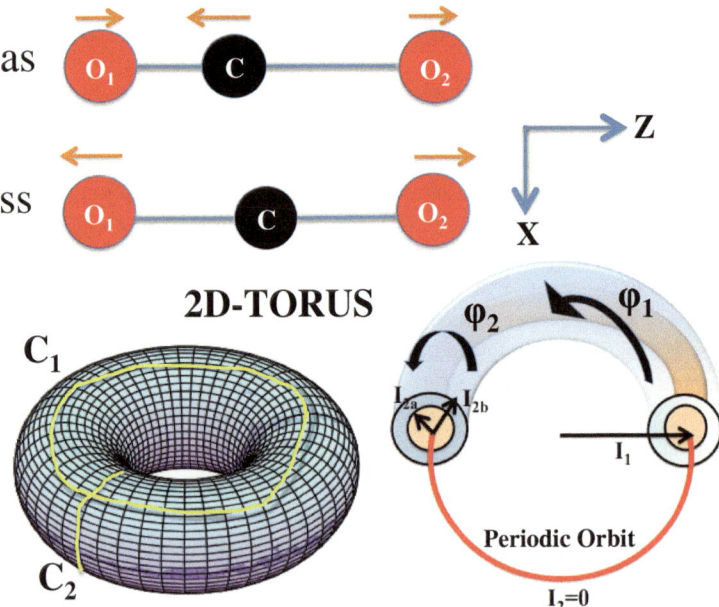

Fig. 1.2 The two vibrational normal modes of carbon dioxide, symmetric stretch (ss) and antisymmetric stretch (as), portrayed in a Cartesian coordinate system, $(\delta Z_{O_1}, \delta Z_C, \delta Z_{O_2})$, with *arrows* and in angle variables (ϕ_1, ϕ_2) of a two-dimensional torus. (I_1, I_2) are the action variables conjugate to the angle variables

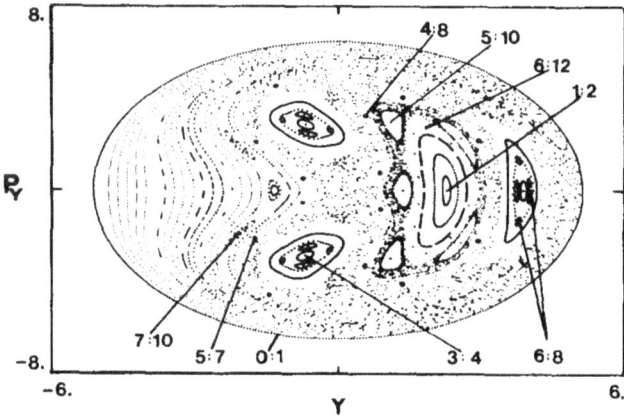

Fig. 1.3 Barbanis-Contopoulos Hamiltonian: $H = \frac{1}{2}(p_x^2 + p_y^2 + \omega_x^2 x^2 + \omega_y^2 y^2) - \varepsilon x^2 y$. The parameters used are $\omega_x^2 = 0.9$, $\omega_y^2 = 1.6$, $\varepsilon = 0.08$ and $\omega_x/\omega_y = 3/4$. The Poincaré surfaces of section for several trajectories on the (y, p_y) plane for $x = 0$ and $p_x > 0$. The sections of several stable periodic orbits are also plotted and labelled. The scattered points among the islands of the 3:4 resonance are the intersections of one chaotic trajectory. All trajectories correspond to total energy of about 21 units [11]

The results of the FPUT model demonstrate this quasiperiodic behaviour. Transition from regular to chaotic dynamics was numerically and systematically studied by the astronomers Hénon and Heiles with a galactic 2D model potential [19]. Nonlinear dynamical systems show the generic behaviour of regular motions at low energies and mixed dynamics, quasiperiodic–chaotic, above some energy. This is best demonstrated by plotting *Poincaré surfaces of sections* (PSS). For conservative 2D Hamiltonians trajectories with different initial conditions but constant total energy lie on a 3D energy surface. By properly choosing a 2D PSS and for bound systems, a trajectory integrated in time crosses this surface repeatedly leaving traces, which may form a smooth curve or scattered points. Figure 1.3 is an example of a PSS obtained with the Barbanis-Contopoulos potential [5, 11]. At fixed energy we expect closed curves for trajectories lying on tori (quasiperiodic), whereas a chaotic trajectory which fills the constant energy 3D hypersurface (volume) has scattered points on the PSS.

In the second half of the twentieth century there was an explosion in theoretical and numerical studies of nonlinear dynamical systems, which have now led to the mature theory of nonlinear mechanics.[2] Unequivocally, the parallel development of computers played a catalytic role for the growth of the theory. Applications of nonlinear mechanics to a diversity of complex systems, from plasmas, atoms and molecules to galaxies and cellular and metabolic networks, have given to the theory

---

[2] The book "Chaos: Classical and Quantum" (http://chaosbook.org/) is a collaborative work to present most of the achievements of nonlinear mechanics mainly obtained in the second half of twentieth century.

a unifying character. By carrying out a nonlinear mechanical analysis one primarily searches for *time invariant structures* in phase or state space, which can explain regularities, rhythms or sudden transitions to irregularities and chaos that are quite often counter-intuitive.

Generally, elementary chemical reactions, dissociation–recombination–isomerization, involve breaking and forming single chemical bonds and most often significant energy is required to overcome potential barriers. Thus, reacting molecules are vibrationally excited species, far from their equilibrium states, rendering the common harmonic normal mode analysis valid at low excitation energies, inaccurate. Potential energy surfaces are nonlinear functions with strong couplings among the degrees of freedom which allow the energy to flow. For this reason, it is not surprising that the advances of nonlinear classical mechanics [14, 30] introduced new methods and concepts in the theories of vibrationally excited and reacting molecules. As found for general nonlinear dynamical systems with a few degrees of freedom, excited molecules are expected to show chaotic motions in which the energy is redistributed to many bonds, resonances among vibrational or vibrational–rotational frequencies, energy localization in specific bonds and bifurcations of vibrational modes to produce new type of motions as energy increases.

At the same time, the advent of lasers in the second half of the twentieth century as a source for coherently exciting molecules in specific modes, as well as the development of molecular beams for a detailed study of chemical reactions, brought new insight in understanding and controlling chemical reactivity at the level of specific atomic and molecular state. Hence, controlling chemical reactions by selecting specific bonds or vibrational excited states of the reactants and analysing the energy disposal into specific vibrational states as well as into rotational and translational degrees of freedom of the product molecules has been an ambitious project in chemical dynamics [25] for a long time. Deuterated methane is a good example for which it has been proved that vibrationally excited states control the course of the reaction with chlorine atoms [7]. Excitation of C–H or C–D overtone states promotes the formation of CH_2D or CH_3 product, respectively. Such selectivity and specificity have been obtained thanks to the progress in experimental spectroscopic techniques [9] and molecular beams. However, this endeavour reveals the problems related to the assignment of the spectra of vibrationally excited molecules and the elucidation of the mechanisms for *intramolecular vibrational energy redistribution* (IVR) the solutions of which require a comprehensive understanding of Molecular Dynamics.

Stimulated emission pumping (SEP) and dispersed fluorescence (DF) spectroscopic methods to excite the molecule at very high vibrational states are ideal to deduce the dynamics close to the isomerization or dissociation threshold. As a matter of fact, the SEP spectra of acetylene were the first which revealed vibrational (quantum) chaos at energies above the threshold for acetylene to vinylidene isomerization [22]. Vibrational overtone spectroscopy has seen similar developments [20]. Crim and coworkers [6] have combined the photoacoustic spectroscopy with a time of flight apparatus to control the products in unimolecular and bimolecular reactions by vibrationally exciting specific chemical bonds of reactant molecules. This bond selective chemistry reveals energy localization in specific bonds. In our days, imaging methods [1] for studying molecular photodissociation and bimolecular collision processes, have successfully been applied to several small polyatomic molecules. Alternative reaction pathways than those anticipated from the PES landscape, such as roaming, have been found by employing this kind of experimental techniques [27].

Since the pioneer work of Karplus et al. in 1965 on the study of $H + H_2$ reaction [23], the classical trajectory method and generally classical mechanics, have been applied to a variety of problems ranging from molecular collisions, interaction of electromagnetic radiation with atoms and molecules and the simulation of macroscopic states of matter [12]. Even problems in molecular physics which are solvable in quantum mechanics are treated by classical mechanics in an effort to achieve a better physical insight. One main reason for the adoption of classical mechanical approximation in chemical dynamics is our ability to perform computations for many body problems, and of course, to obtain results in reasonable agreement with the experiment.

A general argument for justifying applications of classical mechanics to quantum objects such as molecules, has always been the validity of semiclassical theory for heavy particles. At the beginning, the semiclassical quantization rule of Einstein-Brillouin-Keller [8] (EBK) was applied to quantize quasiperiodic trajectories, i.e., for integrable dynamical systems. However, the enthusiasm for a theoretical justification of employing classical mechanics in molecules was quickly dropped, by realizing that molecular systems are generally not integrable, and thus, at high energies most of the trajectories are chaotic for which the EBK semiclassical rule is not valid. The advances of nonlinear classical mechanics which brought a deeper understanding of the *structure of phase space*, i.e., how regular and chaotic trajectories are interwoven in conservative Hamiltonian systems, inevitably raised again the problem of the correspondence of classical to quantum mechanics. Particularly, the field of studying the quantum behaviour of a classically chaotic system has brought much discussion about the meaning of *Quantum Chaos* [15].

Thus, the generic picture of phase space for small polyatomic molecules which has been emerged from all these studies is that, the phase space is predominantly regular at low energies, predominantly chaotic at high energies, and with regular and irregular regions coexisting at intermediate energies. This picture is expected for molecules with one well in the potential energy surface. Most molecules have usually several isotopomers, and this means that the potential energy surfaces have more than one minima separated by saddle points. As we shall see this makes the structure of phase space more complicated. Thus, the classical mechanical study of molecular dynamics inevitably put forward the question:

Do the classical nonlinear mechanical motions in molecules have quantum mechanical counterparts and which are their spectroscopic fingerprints?

This question has given an impetus for the revival of semiclassical theory [4] which formulates the correspondence between quantum eigenfunctions and stationary classical mechanical objects such as *periodic orbits* and tori.

Questions related to energy localization and transfer are currently put forward for biological molecules such as proteins [24]. Time resolved infrared and Raman spectroscopy spanning a time interval from femtoseconds to milliseconds [17], are major spectroscopic techniques for studying the dynamics of biomolecules. Furthermore, efforts to find localized motions in infinite periodic or random anharmonic lattices have led to the concept of *discrete breathers* [3]. The initial theoretical observations of localized motions in the work of Sievers and Takeno [26] triggered the discovery of significant mathematical theorems for the existence of local stationary objects such as periodic orbits in infinite dimensional lattices.

The landscape of the PES may be drastically altered as some parameters in the molecule vary [29]. Barriers and minima may disappear or appear. Similarly, the structure of phase space changes with the total energy. Stable, quasiperiodic motions may turn to unstable chaotic ones and vice versa. But most importantly, new type of motions emanate via *bifurcations* (or branching) of periodic orbits. The bifurcation theory of multidimensional Hamiltonian dynamical systems has mainly been developed in the second half of the twentieth century [18]. One important outcome of the theory is the identification of the *elementary bifurcations* which are described by very simple Hamiltonians as we shall see in Chap. 3. In spite of their simplicity they can also occur in complex dynamical systems at critical energies. This makes elementary bifurcations generic. For molecular Hamiltonian systems we have identify as elementary bifurcations the *center-saddle, period doubling, pitchfork and Hopf.*

Bifurcations are well known in vibrational spectroscopy. For example, the transition from normal to local mode oscillations, first discovered in symmetric ABA molecules, can be understood in classical mechanical phase space as an elementary pitchfork bifurcation [16]. The original local mode models included just the stretching vibrations [21], but they were extended to include bending vibrations as well. Now we know, that the notion of a local mode is more general and it is associated with the bifurcations of classical mechanical stationary objects, such as periodic orbits. Elementary bifurcations are very common in excited polyatomic molecules with the simplest one, the center-saddle, to be ubiquitous. By studying periodic orbits in a parameter space we discover their bifurcations and possible localized eigenstates along them. Periodic orbits which emerge from center-saddle bifurcations appear abruptly at some critical values of the energy, in pairs, and change drastically the geometry of phase space around them. They penetrate in regions of nuclear phase space which the normal mode motions can not reach. Center-saddle bifurcations are of generic type, i.e., they are robust and persist for small (perturbative) changes of the potential function.

1.2 Hierarchical Study of Nonlinear Molecular Dynamics

Nonlinear mechanics offer a systematic way to study complex systems. By investigating the dynamics, what exactly we want is to locate invariant structures in phase space where trajectories live for a long time. Molecular Hamiltonians are usually written as the sum of the potential energy function and kinetic energy. Thus, the hierarchical detailed exploration of the molecular phase space structure, one of the

Fig. 1.4 Hierarchical
investigation of the phase
space structure of a molecule

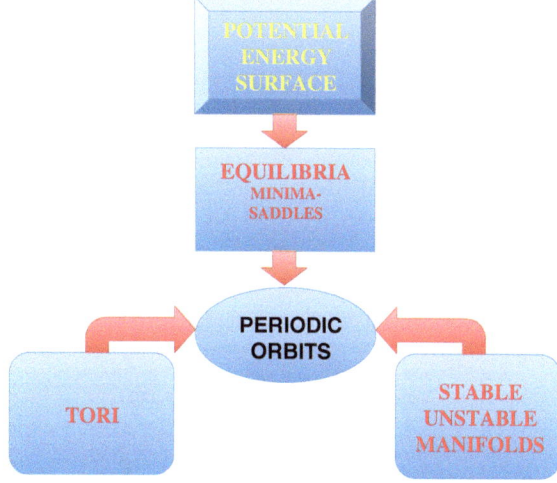

main subjects of this book, requires, first to locate the stationary points of the potential function, second to find periodic orbits (PO) that emanate from the equilibria,
and then to describe tori around stable PO as well as the stable and unstable manifolds of unstable PO. This strategy is graphically illustrated with Fig. 1.4. Saddles on
the PES are unstable equilibria and associated with them there are *normally hyperbolic invariant manifolds* (NHIM), which have been proved to provide an accurate
definition of the transition state in chemical reactions of polyatomic molecules [28].

A global molecular Hamiltonian describes all isomers and dissociation channels
accessible in the energy interval of interest. We use such a Hamiltonian to locate
equilibrium points and periodic orbits. Further phase space structures around equilibrium points and periodic orbits are obtained by taking a Taylor expansion of the
global Hamiltonian up to the order which satisfies a predetermined accuracy. The
quadratic part of the Taylor expansion is used to calculate the normal coordinates,
whereas higher order terms in the Taylor expansion of the global Hamiltonian are
used to find the normal form coordinates. The latter allows one to construct an
approximate integrable Hamiltonian, valid however, for energies near to equilibrium
point. The integrals of motion of this integrable Hamiltonian (action variables) can
semiclassically be quantized according to EBK rules [8], provided there is no resonance condition. The latter means that for a n degrees of freedom system there is no
relation $\sum_{i=1}^{n} m_i \omega_i = 0$, where m_i are integer numbers and ω_i the fundamental frequencies of the n oscillators. Quantum mechanical calculations are straightforward
in normal form coordinates. In the following chapters we introduce and explain the
above phase space structures and describe methods for their numerical calculation.

References

1. Ashfold MNR, Nahler NH, Orr-Ewing AJ, Vieuxmaire OPJ, Toomes RL, Kitsopoulos TN, Garcia IA, Chestakov DA, Wu SM, Parker DH (2006) Imaging the dynamics of gas phase reactions. PCCP 8:26–53

2. Born M, Oppenheimer JR (1927) On the quantum theory of molecules. Ann Physik 84:457–484. Translated by S. M. Blinder, http://en.wikipedia.org/wiki/Born-Oppenheimer_approximation

3. Campbell DK, Flach S, Kivshar YS (2004) Localizing energy through nonlinearity and discreteness. Phys Today 43(1):43–49

4. Child MS (1991) Semiclassical mechanics with molecular applications. Oxford University Press, US

5. Contopoulos G (1970) Orbits in highly perturbed dynamical systems. I Periodic orbits Astron J 75:96–107

6. Crim FF (1999) Vibrational state control of bimolecular reactions: discovering and directing the chemistry. Acc Chem Res 32(10):877–884

7. Crim FF (2008) Chemical dynamics of vibrationally excited molecules: controling reactions in gases and on surfaces. In: Proceedings of National Academy of Science, US, vol 105, pp 12,654–12,661

8. Curtis LJ, Ellis DG (2004) Use of the Einstein-Brillouin-Keller action quantization. Am J Phys 72(12):1521–1523

9. Dai HL, Field R (1995) Molecular dynamics and spectroscopy by stimulated emission pumping., Advanced series in physical chemistryWorld Scientific Publishing, Singapore

10. Ford J (1992) The Fermi-Pasta-Ulam problem: paradox turns discuvery. Phys Rep 213(5):271–310

11. Founargiotakis M, Farantos SC, Contopoulos G, Polymilis C (1989) Periodic orbits, bifurcations and quantum mechanical eigenfunctions and spectra. J Chem Phys 91(1):1389–1402

12. Frenkel D, Smit B (1996) Understanding molecular simulation. Academic Press, New York

13. Goldstein H (1977) Classical mechanics, twelth edn. Addison-Wesley Publishing, UK

14. Guckenheimer J, Holmes P (1990) Nonlinear oscillations, dynamical systems, and bifurcations of vector fields, 2nd edn., Applied mathematical sciences, Springer, New York

15. Gutzwiller MC (1990) Chaos in classical and quantum mechanics. Springer, New York

16. Halonen L (1998) Local mode vibrations in polyatomic molecules. Adv Chem Phys 104(12):41–179

17. Hamm P, Helbing J, Bredenbeck J (2008) Two-dimensional infrared spectroscopy of photoswitchable peptides. Annu Rev Phys Chem 59:291–317

18. Hanssmann H (2007) Local and semi-local bifurcations in Hamiltonian dynamical systems: results and examples. Springer, Heidelberg

19. Hénon M, Heiles C (1964) The applicability of the third integral of motion: some numerical experiments. Astron J 69:73–79

20. Henry BR (1993) Vibrational spectra and structure, vol 10. Elsevier, New York

21. Herzberg G (1945) Infrared and Raman spectra. Van Nostrand, New York

22. Jonas DM, Solina SAB, Rajaram B, Cohen SJ, Field RW, Yamanouchi K, Tsuchiya S (1993) Intramolecular vibrational redistribution of energy in the stimulated emission pumping spectrum of acetylene. J Chem Phys 99:7350–7370

23. Karplus M, Porter RN, Sharma RD (1965) Exchange reactions with activation energy: I. Simple barrier potential for (H, H_2). J Chem Phys 43:3259

24. Leitner DM (2008) Energy flow in proteins. Annu Rev Phys Chem 59:233–259

25. Levine RD, Bernstein RB (1987) Molecular reaction dynamics and chemical reactivity, 3rd edn. Oxford University Press, New York

26. Sievers AJ, Takeno S (1988) Intrinsic localized modes in anharmonic crystals. Phys Rev Lett 61(8):970–973

27. Suits AG (2008) Roaming atoms and radicals: a new mechanism in molecular dissociation. Acc Chem Res 41(7):873–881

28. Waalkens H, Schubert R, Wiggins S (2008) Wigner's dynamical transition state theory in phase space: classical and quantum. Nonlinearity 21:R1–R118
29. Wales DJ (2004) Energy landscapes: applications to clusters, biomolecules and glasses. Cambridge University Press, Cambridge
30. Wiggins S (2003) Introduction to applied nonlinear dynamical systems and chaos, 2nd edn. Springer, New York

Chapter 2
The Geometry of Hamiltonian Mechanics

In this chapter an introduction to Hamiltonian mechanics is given. Although, most of the textbooks devote one or more chapters to the Hamiltonian formulation of classical mechanics, only a few approach the subject from the theory of differential geometry [1, 3, 5]. The latter neatly exposes the geometrical properties of Hamiltonian mechanics. Modern analysis on manifolds [7] provides the means to develop the theory in a coordinate free way. However, numerical applications require the translation of the theory to specific coordinate systems. Hence, in this introductory chapter we follow both approaches to unveil the geometrical properties of Hamiltonian mechanics [4, 6]. This chapter must be read in parallel with the Appendix where some basic definitions and theorems from the calculus on manifolds are provided.

2.1 Configuration Manifolds and Coordinate Systems

2.1.1 Cartesian Coordinates

We consider a system of N particles whose configurations in a space fixed Cartesian coordinate system are described by N vectors of three components or with single vectors of $3N$ components. The Cartesian configuration space consists an **Euclidean manifold** (M) of dimension $3N$, $M \subset \mathbb{R}^{3N}$. The number of *degrees of freedom* for the system is $3N$. The positions of N particles with masses $m_\alpha, \alpha = 1, \ldots, N$, in our 3D world, are described by N vectors r^α

$$r^\alpha = x^\alpha \mathbf{i} + y^\alpha \mathbf{j} + z^\alpha \mathbf{k}, \quad \alpha = 1, \ldots, N. \tag{2.1}$$

$(\mathbf{i}, \mathbf{j}, \mathbf{k})$ denote the unit vectors along the (x, y, z)-axes, respectively.

The mechanical state of the system is defined by the coordinates of the particles and the rate of their change in time t, the velocities;

© The Author(s) 2014
S.C. Farantos, *Nonlinear Hamiltonian Mechanics Applied to Molecular Dynamics*,
SpringerBriefs in Electrical and Magnetic Properties of Atoms, Molecules, and Clusters,
DOI 10.1007/978-3-319-09988-0_2

$$v^\alpha = \frac{dr^\alpha}{dt} \equiv \dot{r}^\alpha = \frac{dx^\alpha}{dt}\mathbf{i} + \frac{dy^\alpha}{dt}\mathbf{j} + \frac{dz^\alpha}{dt}\mathbf{k}$$
$$\equiv \dot{x}^\alpha\mathbf{i} + \dot{y}^\alpha\mathbf{j} + \dot{z}^\alpha\mathbf{k}, \quad \alpha = 1, \ldots, N. \tag{2.2}$$

Thus, the time evolution of the system is completely determined by the vectors, $[r^\alpha(t), v^\alpha(t)]$, $\alpha = 1, \ldots, N$.

The *kinetic energy* of the N-particle system is defined by the quadratic function in velocities

$$K = \frac{1}{2}\sum_{\alpha=1}^{N} m_\alpha (v^\alpha)^2$$
$$= \frac{1}{2}\sum_{\alpha=1}^{N} m_\alpha \left[(\dot{x}^\alpha)^2 + (\dot{y}^\alpha)^2 + (\dot{z}^\alpha)^2\right]. \tag{2.3}$$

The interactions among the particles are determined by the *potential energy*, $V(r^1, \ldots, r^N)$, i.e., a function of the position vectors. The resultant *force* on particle α is the vector

$$F_\alpha = -\frac{\partial V(r)}{\partial r^\alpha} \equiv -\partial_\alpha V(r)$$
$$= -\frac{\partial V}{\partial x^\alpha}\mathbf{i} - \frac{\partial V}{\partial y^\alpha}\mathbf{j} - \frac{\partial V}{\partial z^\alpha}\mathbf{k}$$
$$= F_{x_\alpha}\mathbf{i} + F_{y_\alpha}\mathbf{j} + F_{z_\alpha}\mathbf{k}. \tag{2.4}$$

2.1.2 Curvilinear Coordinates

Because of some geometrical constraints or space-time symmetries which result in conservation laws, such as of the total energy, momentum and angular momentum, and the possible existence of other constants (integrals) of motion, the orbits of the particles are constrained in a configuration space with dimension less than $3N$. If there are k holonomic constraint equations[1]

$$\phi^i(r^1, \ldots, r^N) = c^i, \quad i = 1, \ldots, k, \tag{2.5}$$

that assign specific values to the associated quantities, geometrical or constants of motion, then, the number of degrees of freedom is $n = 3N - k$, and the configurations of the system form a ***smooth (differentiable) manifold*** Q of dimension n (see Appendix A), not necessarily Euclidean. The k constraint equations provide an implicit representation of the ***configuration manifold*** (see Appendix A.2).

---

[1] Holonomic constraints may contain the velocities $\phi^i(r^1, \ldots, r^n, \dot{r}^1, \ldots, \dot{r}^N) = c^i$, which however, can be integrated to equations without the velocities.

Taking into account possible constraint equations we may want to study the orbits of the system on the reduced dimension, n, configuration manifold Q. This may be an imperative step for extracting the underlying physics out of the dynamics of the system. Smooth manifolds can be covered by **atlases of charts**, which locally define maps of open sets of the manifold to open sets of an Euclidean space. In this way we introduce generalized coordinates, $(q^1, \ldots, q^n)$, and apply ordinary calculus to study the dynamics of the system. However, it is worth emphasizing that global properties of manifolds may be studied without any reference to a local coordinate system. In principle and with the aid of the k constraint equations, one can find transformation equations from the n generalized coordinates to the $3N = n + k$ Cartesian coordinates

$$
\begin{aligned}
x^\alpha &= g_{x^\alpha}(q^1, \ldots, q^n, c^1, \ldots, c^k) \\
y^\alpha &= g_{y^\alpha}(q^1, \ldots, q^n, c^1, \ldots, c^k) \\
z^\alpha &= g_{z^\alpha}(q^1, \ldots, q^n, c^1, \ldots, c^k), \quad \alpha = 1, \ldots, N.
\end{aligned}
\tag{2.6}
$$

The generalized velocities $(\dot{q}^1, \ldots, \dot{q}^n)$ are related to Cartesian velocities by the equations,

$$
\dot{x}^\alpha = \sum_{k=1}^{n} \frac{\partial g_{x^\alpha}}{\partial q^k} \dot{q}^k
$$

$$
\dot{y}^\alpha = \sum_{k=1}^{n} \frac{\partial g_{y^\alpha}}{\partial q^k} \dot{q}^k
$$

$$
\dot{z}^\alpha = \sum_{k=1}^{n} \frac{\partial g_{z^\alpha}}{\partial q^k} \dot{q}^k, \quad \alpha = 1, \ldots, N.
\tag{2.7}
$$

Then, the kinetic energy (Eq. 2.3) in generalized coordinates will take the form,

$$
K = \frac{1}{2} \sum_{i,k=1}^{n} \dot{q}^i g_{ik}(q, m) \dot{q}^k,
\tag{2.8}
$$

where, $g_{ik}(q, m)$ is the **metric tensor** and its components are functions of the masses, $m = (m_1, \ldots, m_\alpha, \ldots, m_N)$, and generalized coordinates, $q = (q^1, \ldots, q^n)^T$ [2] (see next section). The sum of kinetic and potential energy is the *total energy* of the system

$$
E[q(t), \dot{q}(t)] = K[q(t), \dot{q}(t)] + V[q(t)].
\tag{2.9}
$$

We must admit that the transformation equations from curvilinear to Cartesian coordinates and their inverses (Eq. 2.6) are not always easy to find. As a matter of

---

[2] The letter superscript (T) denotes a column vector and generally the transpose of a matrix.

fact, to determine the constants of motion for a dynamical system requires one to know the solutions of the equations of motion. The equations of motion take a simple form in Cartesian coordinates and can be solved numerically with modern computers for large systems with thousand of atoms. Combining, integration of equations of motion in Cartesian coordinates and transforming to specific curvilinear coordinates to describe the manifolds on which the trajectories lie is an appealing approach to illuminate Molecular Dynamics.

2.2 The Topological Map of Lagrangian and Hamiltonian Mechanics

Topological theories by not relying on specific coordinate systems have the advantage to reveal the general geometrical properties of physical systems, and thus, they are suitable for a qualitative analysis. In reverse, by knowing the topological structure of the system one can choose a suitable local coordinate system for computational work. Figure 2.1 portrays the topological structures of the two main formulations of Classical Mechanics, the Lagrangian and Hamiltonian. By considering the configuration space of a dynamical system as a smooth (differentiable) manifold, Q, there is always a **chart** (a local coordinate system), i.e., a homeomorphism (see Appendix A),

$$\phi : U \subset Q \to \phi(U) \subset \mathbb{R}^n, \tag{2.10}$$

of an open set U of Q onto an open set $\phi(U)$ of $\mathbb{R}^n$. Since, the map is on an Euclidean space ($\mathbb{R}^n$), we can also define a coordinate representation in $\mathbb{R}^n$

$$q^i = f^i \circ \phi \quad \text{or} \quad \phi(s) = (q^1(s), q^2(s), \dots, q^n(s))^T \in \mathbb{R}^n, \tag{2.11}$$

for every point $s \in U$, and f^i are differentiable functions. The **tangent space** of Q (the space where the derivatives live) at a point $s \in Q$ ($T_s Q$) is a vector space (velocities belong to this space) and the union of all tangent spaces for all points s of Q form the **tangent bundle** (TQ) with Q the **base space**

$$TQ = \bigcup_{s \in Q} T_s Q. \tag{2.12}$$

The tangent bundle contains both the manifold Q and its tangent spaces $T_s Q$ called the **fibres** and it is a smooth manifold of dimension $2n$. Since, TQ is also a smooth manifold a chart can be defined by the diffeomorphism

$$T\phi : TU \to \phi(U) \times \mathbb{R}^n \subset \mathbb{R}^n \times \mathbb{R}^n. \tag{2.13}$$

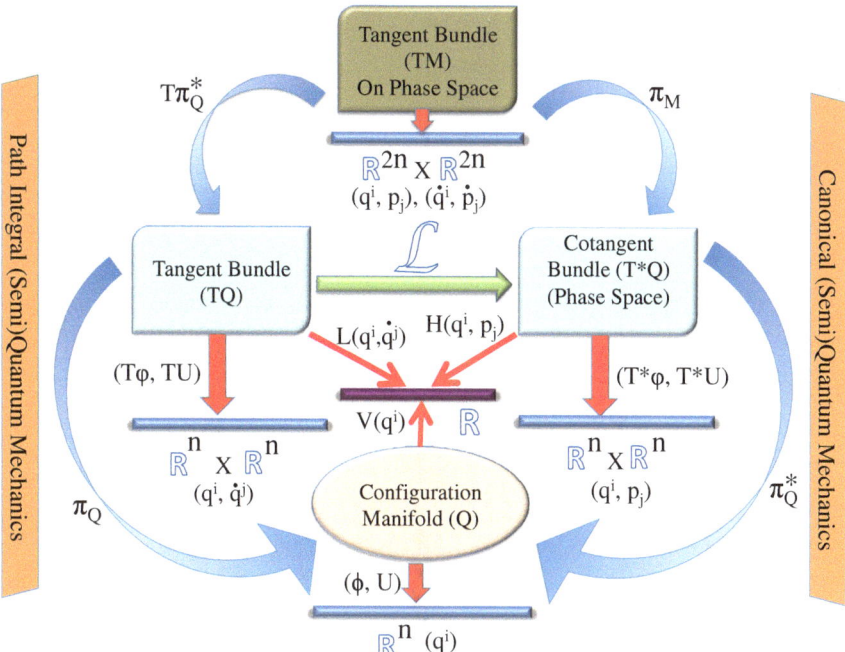

Fig. 2.1 Topological map of Lagrangian and Hamiltonian Mechanics. The tangent bundle (TQ) of the configuration manifold (Q) is a smooth manifold with charts defined by the generalized coordinates (q^i) and their corresponding velocities $(\dot{q}^j)$. The Lagrangian, $L[q(t), \dot{q}(t)]$, is a function on the tangent bundle to real numbers. The dual space of TQ is the cotangent bundle $(M = T^*Q)$, also named phase space. The phase space is a differentiable manifold of dimension $2n$ for which the tangent bundle, $TM \equiv T(T^*Q)$ of dimension $(2n \times 2n)$, can also be defined with charts described by the generalized coordinates (q^i), the conjugate momenta (p_j) and their velocities $(\dot{q}^i, \dot{p}_j)$. The potential function, $V(q)$, is a function on the configuration manifold to real numbers. The Hamiltonian, $H[q(t), p(t)]$, is a function on the phase space to real numbers obtained by a Legendre transform $(\mathbb{L})$ of the Lagrangian. We may consider that the Legendre transform generates a differentiable map between the tangent and cotangent bundles of Q, $F_{\mathbb{L}} : TQ \to T^*Q$. Then, the tangent mapping $TF_{\mathbb{L}}$ defines an isomorphism between the tangent of tangent bundle of Q (not shown) and the tangent bundle of phase space, $TF_{\mathbb{L}} : T(TQ) \to T(T^*Q)$. π_Q, π_Q^* and π_M are canonical projections. $T\pi_Q^*$ is the tangent mapping of π_Q^*. In Chap. 4 we discuss how the Lagrange formalism of classical mechanics leads to the path integral formulation of quantum mechanics and the Hamiltonian mechanics to canonical quantum mechanics

This is a linear map and each chart (ϕ, U) from the atlas of Q induces a chart $(T\phi, TU)$ for TQ. This chart is said to be the **bundle chart** associated with (ϕ, U).

   The potential function V is a map of configuration manifold to real numbers $\mathbb{R}$, $V : Q \to \mathbb{R}$. On the tangent bundle we define the *state function*

$$L : TQ \to \mathbb{R}, \tag{2.14}$$

named *Lagrangian*. Having defined a chart the Lagrangian takes the form

$$L(q, \dot{q}) = K(q, \dot{q}) - V(q). \tag{2.15}$$

By using the Lagrangian we define the generalized momenta as

$$p_i(q, \dot{q}) = \frac{\partial L}{\partial \dot{q}^i}. \tag{2.16}$$

To extract the physical meaning of these derivatives we write the Lagrangian in Cartesian coordinates, Eq. 2.3.

$$L = K - V = \frac{1}{2} \sum_{\alpha=1}^{N} m_\alpha (v^\alpha)^2 - V(r), \quad r = (r^1, \dots, r^N). \tag{2.17}$$

The partial derivative of L with respect to the position vector of particle α, r_α, is the force acting on this particle, Eq. (2.4), whereas the partial derivative with respect to the velocity of particle α is

$$\frac{\partial L}{\partial v^\alpha} = m_\alpha v^\alpha. \tag{2.18}$$

The vector quantity

$$p_\alpha = m_\alpha v^\alpha = m_\alpha (\dot{x}^\alpha \mathbf{i} + \dot{y}^\alpha \mathbf{j} + \dot{z}^\alpha \mathbf{k}), \tag{2.19}$$

is the *momentum* of particle α.

Writing the Lagrangian in generalized coordinates,

$$L(q, \dot{q}) = \frac{1}{2} \sum_{i,j=1}^{n} \dot{q}^i g_{ij}(q, m) \dot{q}^j - V(q), \tag{2.20}$$

we define the component of the generalized force along the ith degree of freedom as

$$f_i = \frac{\partial L}{\partial q^i}, \tag{2.21}$$

and the component of the generalized momentum along the ith degree of freedom

$$p_i = \frac{\partial L}{\partial \dot{q}^i} = \sum_j g_{ij} \dot{q}^j. \tag{2.22}$$

The *tangent* and *cotangent bundle* (see Appendix A.8), TQ and T^*Q respectively, exist for any configuration manifold Q. If, however, we can define a metric on the manifold, i.e., Q is a Riemannian manifold, then, there is a diffeomorphism

$TQ \to T^*Q$ that sends the coordinate patch $(q, \dot{q})$ on the tangent space at a point s of Q to the coordinate patch (q, p) on the **cotangent space**. Taking as a metric the covariant tensor rank-2, g_{ij}, that defines the kinetic energy, then, the momentum p_i is the **covector** of the velocity $\dot{q}^i$, and the velocity $\dot{q}^i$ can be obtained by the inverse tensor g^{ij}

$$\dot{q}^i = \sum_j g^{ij} p_j, \qquad (2.23)$$

where $\sum_l g_{il} g^{lj} = \delta_i^j$.[3]

The metric tensor is a 2−**form** (see A.8), and thus, acts on two vectors of the tangent space to map them to a scalar. In other words, we can write

$$g_s(v, w) = \sum_{i=1}^n \sum_{j=1}^n v^i g_{ij} w^j, \qquad (2.24)$$

where v^i and w^j are the components of the two vectors v and w of the tangent space, $T_s Q$, at the point s of the manifold Q, respectively in a local coordinate system. In a coordinate free interpretation of the metric, the kinetic energy is just the half of the metric, $K = \frac{1}{2} g_s(v, v)$. We may also consider the metric g_s to act only on one vector field, a mapping from TQ to T^*Q, i.e.,

$$g_s : TQ \to T^*Q : v \mapsto g_s(\bullet, v), \qquad (2.25)$$

with $\bullet$ to denote a vacancy in the pair of vectors. Thus, $g_s(\bullet, v)$ is a 1−**form**, which can act on another or the same vector in $T_s Q$ to yield a real number, $g_s(v, v)$. The metric assigns to each **vector field** $X \in \mathscr{X}(Q)$ the smooth 1−form $g(\bullet, X) \in \mathscr{X}^*(Q)$, and vice versa. $\mathscr{X}(Q)$ is the set of vector fields on the configuration manifold Q and $\mathscr{X}^*(Q)$ the set of covectors. Therefore, we may conclude that, in charts the generalized momenta p_i, which is canonically conjugate to the coordinates q^i, is the 1−form

$$p_i = \frac{\partial L}{\partial \dot{q}^i} = \sum_j g_{ij} \dot{q}^j \equiv g_s(\bullet, v), \qquad (2.26)$$

which is a map from the tangent bundle (TQ) to the cotangent bundle (T^*Q). In fact, $(q, p) \equiv (q^1, \ldots, q^n, p_1, \ldots, p_n)$ are the local coordinates in the cotangent bundle which is called the *phase space* of the dynamical system.

The *Hamiltonian*, $H(q^i, p_j)$, is a function on the phase space to real numbers obtained by a *Legendre transform* ($\mathbb{L}$) of the Lagrangian

---

[3] The components of Kronecker delta tensor, δ_i^j, are equal to 1 for $i = j$ and 0 for $i \neq j$.

$$H(q, p) = \sum_{i=1}^{n} \dot{q}^i p_i - L(q, \dot{q}) = \sum_{i=1}^{n} \dot{q}^i \sum_{j=1}^{n} g_{ij}\dot{q}^j - L(q, \dot{q})$$

$$= \frac{1}{2} \sum_{ij} \dot{q}^i g_{ij}\dot{q}^j + V(q) = K + V. \tag{2.27}$$

The transformation equations in a new coordinate system in the configuration space lead to the following transformation equations for the velocities in the tangent space

$$Q^i = Q^i(q^1, \ldots, q^n), \quad i = 1, \ldots, n \tag{2.28}$$

$$\dot{q}^j = \sum_i \left(\frac{\partial q^j}{\partial Q^i} \right) \dot{Q}^i, \tag{2.29}$$

and the new momenta in the cotangent space

$$P_i = \frac{\partial L}{\partial \dot{Q}^i} = \sum_j \left(\frac{\partial L}{\partial \dot{q}^j} \right) \left(\frac{\partial \dot{q}^j}{\partial \dot{Q}^i} \right)$$

$$= \sum_j p_j \left(\frac{\partial q^j}{\partial Q^i} \right). \tag{2.30}$$

2.3 The Principle of Least Action

The function

$$S(q_a, q_b; t_a, t_b) = \int_{t_a}^{t_b} L[q(t), \dot{q}(t)]dt, \tag{2.31}$$

is called the *action* along the path that connects the configuration points q_a and q_b at the times t_a and t_b, respectively;

$$q_a = q(t_a), \quad q_b = q(t_b). \tag{2.32}$$

In mechanics we accept the *Principle of Least Action*; among the infinite number of paths between two fixed configuration points (q_a, q_b) and times t_a and t_b the system will follow that one which minimizes the action (Eq. 2.31),

$$S_0 = min\,[S(q_a, q_b; t_a, t_b)] = min \int_{t_a}^{t_b} L[q(t) + \delta q(t), \dot{q}(t) + \delta \dot{q}(t)]dt. \tag{2.33}$$

We assume the two end points fixed and we expand S in δq. The variation of S around an *extremum* q is

$$\delta S = \int_{t_a}^{t_b} L[q(t) + \delta q(t), \dot{q}(t) + \delta \dot{q}(t)]dt - \int_{t_a}^{t_b} L[q(t), \dot{q}(t)]dt$$

$$= \int_{t_a}^{t_b} \left(\frac{\partial L}{\partial q} \delta q + \frac{\partial L}{\partial \dot{q}} \delta \dot{q} \right) dt = 0. \tag{2.34}$$

Integrating by parts we have

$$\delta S = \int_{t_a}^{t_b} \left(\frac{\partial L}{\partial q} \delta q + \frac{d}{dt}\left[\frac{\partial L}{\partial \dot{q}} \delta q \right] - \frac{d}{dt}\left[\frac{\partial L}{\partial \dot{q}} \right] \delta q \right) dt$$

$$= \left[\frac{\partial L}{\partial \dot{q}} \delta q \right]_{t_a}^{t_b} + \int_{t_a}^{t_b} \left(\frac{\partial L}{\partial q} - \frac{d}{dt}\frac{\partial L}{\partial \dot{q}} \right) \delta q\, dt = 0. \tag{2.35}$$

Since, $\delta q_a = \delta q_b = 0$ and δS is zero for any positive or negative variation of δq, we infer that

$$\frac{\partial L}{\partial q} - \frac{d}{dt}\frac{\partial L}{\partial \dot{q}} = 0. \tag{2.36}$$

These are the *Euler–Lagrange equations*. Hence, according to the variational principle the equations of motion define the path for which the action takes a stationary value. Generally, for a system with n degrees of freedom is valid

$$\frac{\partial L}{\partial q^i} - \frac{d}{dt}\frac{\partial L}{\partial \dot{q}^i} = 0, \quad i = 1, \ldots, n. \tag{2.37}$$

For a Lagrangian written as in Eq. 2.17 the Euler–Lagrange equation (Eq. 2.37) takes the form

$$f_i = \dot{p}_i, \quad i = 1, \ldots, n, \tag{2.38}$$

i.e., *Newton's equations*. The importance of the Lagrangian stems from its utility to define the action along a path between two configuration points, (q_a, q_b). Hence, the action is a function of the initial and final configuration points as well as the time. The principle of the least action leads to the equations of motion, which involve the partial derivatives of the Lagrangian defined on the tangent space, TQ.

2.4 Hamiltonian Vector Fields

In Sect. 2.2 we introduced the Hamiltonian state function in phase space, the cotangent space of the tangent space of the configuration manifold. As we shall see, the Hamiltonian formalism of classical mechanics is the most appropriate to reveal intrinsic symmetries of the system, and the entrance to quantum and statistical mechanics. Thus, it is worth formulating classical mechanics in phase space. The Hamiltonian of a mechanical system in phase space, $H(q, p, t)$, is a function of coordinates, momenta and possibly of time. Then, the equations of motion can be inferred from the Principle of Least Action

$$\delta S(q_a, q_b; t_a, t_b) = \delta \left(\int_{t_a}^{t_b} \left(\sum_{i=1}^{n} p_i \dot{q}^i - H \right) dt \right) = 0, \tag{2.39}$$

with fixed end points. This equation is transformed to

$$\delta S = \sum_{i=1}^{n} \int_{t_a}^{t_b} \left(\delta p_i \dot{q}^i + p_i \delta \dot{q}^i - \frac{\partial H}{\partial q^i} \delta q^i - \frac{\partial H}{\partial p_i} \delta p_i \right) dt$$

$$= \sum_{i=1}^{n} \left[\int_{t_a}^{t_b} \delta p_i \left(\dot{q}^i - \frac{\partial H}{\partial p_i} \right) dt - \int_{t_a}^{t_b} \delta q^i \left(\dot{p}_i + \frac{\partial H}{\partial q^i} \right) dt + \left[p_i \delta q^i \right]_{t_a}^{t_b} \right]$$

$$= 0. \tag{2.40}$$

The last term evaluated at the end points is zero and the independent variations of δq_i and δp_i lead to *Hamilton's equations*

$$\dot{q}^i = \frac{\partial H}{\partial p_i}$$

$$\dot{p}_i = -\frac{\partial H}{\partial q^i}, \quad i = 1, \ldots, n. \tag{2.41}$$

Equations 2.41 define the local **flow** of the *Hamiltonian vector field*. If we denote this vector field as $(X_H, \overline{(X_H)})^T$ to distinguish coordinates from momenta, then, the Hamiltonian vector field in local coordinates is written as

$$\begin{pmatrix} (X_H)^i \\ \overline{(X_H)_j} \end{pmatrix} = \begin{pmatrix} \partial H / \partial p_i \\ -\partial H / \partial q^j \end{pmatrix}, \quad i, j = 1, \ldots, n. \tag{2.42}$$

Hence, the principle of least action results in the Euler–Lagrange equations in the Lagrangian formalism, whereas in the Hamiltonian formalism of classical mechanics it gives Hamilton's equations. However, it is important to understand that in the

Lagrangian formalism the dynamics take place in the tangent of the tangent bundle of configuration manifold, $T(TQ)$, and in the Hamiltonian formalism in the tangent of the cotangent bundle of configuration manifold, $T(T^*Q)$. Since, the Lagrangian and Hamiltonian state functions are connected by a Legendre transform it can be proved that the two formulations of classical mechanics are equivalent.

Finally, if we consider the action as a function of the initial and final coordinates, not fixed but taking the path that minimizes the action, i.e., the generalized coordinates describe an integral solution of Hamilton's equations, then, the variation of the action is

$$
\begin{aligned}
\delta S(q_a, q_b) &= \sum_{i=1}^{n} \int_{t_a}^{t_b} \left(\delta p_i \dot{q}^i + p_i \delta \dot{q}^i - \frac{\partial H}{\partial q^i} \delta q^i - \frac{\partial H}{\partial p_i} \delta p_i \right) dt \\
&= \sum_{i=1}^{n} \left[\int_{t_a}^{t_b} \delta p_i \left(\dot{q}^i - \frac{\partial H}{\partial p_i} \right) dt - \int_{t_a}^{t_b} \delta q^i \left(\dot{p}_i + \frac{\partial H}{\partial q^i} \right) dt + \left[p_i \delta q^i \right]_{t_a}^{t_b} \right] \\
&= \sum_{i=1}^{n} \left[p_{ib} \delta q_b^i - p_{ia} \delta q_a^i \right],
\end{aligned}
\tag{2.43}
$$

where a and b denote the end points of the path. Thus,

$$
\begin{aligned}
\frac{\partial S}{\partial q_a^i} &= -p_{ia} \\
\frac{\partial S}{\partial q_b^i} &= p_{ib}, \quad i = 1, \dots, n.
\end{aligned}
\tag{2.44}
$$

Similarly, if we consider the action as a function of the coordinates and time

$$
\begin{aligned}
\frac{d}{dt} S(q, t) = L &= \frac{\partial S}{\partial t} + \frac{\partial S}{\partial q} \dot{q} \\
&= \frac{\partial S}{\partial t} + p\dot{q}.
\end{aligned}
\tag{2.45}
$$

Hence,

$$
\frac{\partial S}{\partial t} = L - p\dot{q} = -H.
\tag{2.46}
$$

From the above equations we can write the total differential of action as

$$
dS(q, t) = \sum_{i=1}^{n} p_i dq^i - H(q, p, t) dt.
\tag{2.47}
$$

2.5 The Canonical Equations Expressed with the Symplectic 2−Form

By replacing velocities with momenta not only second order differential equations (Euler–Lagrange) are replaced by the first order equations of Hamilton, but as we shall see, generalized coordinates and their conjugate momenta acquire equivalent significance and reveal the geometry of phase space. Let us first collect the generalized coordinates and their conjugate momenta of a dynamical system of n degrees of freedom to a single vector $x = (q^1, q^2, \ldots, q^n, p_1, p_2, \ldots, p_n)^T$ of $2n$-dimension. Then, Hamilton's equations are written in the form

$$\dot{x}(t) = J \partial H(x), \tag{2.48}$$

where ∂H is the gradient of Hamiltonian function, and J the *symplectic matrix*

$$J = \begin{pmatrix} 0_n & I_n \\ -I_n & 0_n \end{pmatrix}. \tag{2.49}$$

0_n and I_n are the zero and unit $n \times n$ matrices, respectively. It is proved that J satisfies the relations,

$$J^{-1} = -J = J^T \quad \text{and} \quad J^2 = -I_{2n}. \tag{2.50}$$

$X_{H_x} = J \partial H(x)$ is the Hamiltonian vector field as was defined by Eq. 2.42. In fact, as was discussed in Sect. 2.2 and from Fig. 2.1 (top) we can infer that x defines a chart in the tangent space (TM) of phase space M.

Let us denote with θ the 1−forms defined on the phase space manifold M

$$\theta : M \to T^*M : m \in M \mapsto \theta_m \in T^*_m M, \tag{2.51}$$

and with α the 1−forms on the configuration manifold Q

$$\alpha : Q \to T^*Q : r \in Q \mapsto \alpha_r \in T^*_r Q. \tag{2.52}$$

Since, α is a linear map from Q to M and θ an 1−form on M we can **pull-back** θ to Q to produce the 1−form $\alpha^*\theta$, which lives on the base manifold Q. Then, the **Canonical Poincaré 1−Form** is given by

$$\hat{\theta} = \sum_i p_i dq^i, \tag{2.53}$$

and satisfies the relation

$$\alpha^*\hat{\theta} = \alpha \quad \text{forall} \ \alpha \in \mathcal{X}^*(Q). \tag{2.54}$$

$\hat{\theta}$ is invariant under coordinate transformations. This is proved by using Eq. 2.30. Indeed,

$$\hat{\theta} = \sum_i p_i dq^i = \sum_i p_i \sum_j \frac{\partial q^i}{\partial Q^j} dQ^j$$

$$= \sum_j \left(\sum_i p_i \frac{\partial q^i}{\partial Q^j} \right) dQ^j = \sum_j P_j dQ^j. \tag{2.55}$$

The **Canonical Symplectic 2− Form** is extracted by taking the *exterior derivative* of $\hat{\theta}$

$$\hat{\omega} \equiv \overset{2}{\omega} = -d\hat{\theta}. \tag{2.56}$$

This is a *closed* 2−*form* ($d\hat{\omega} = -d \circ d\hat{\theta} = 0$). In local coordinates (q, p), $\hat{\omega}$ is expressed by the *wedge products* (Darboux's theorem)

$$\hat{\omega}_r = \sum_i dq^i \wedge dp_i, \quad r \in M. \tag{2.57}$$

If we introduce $dx = (dq^1, \ldots, dq^n, dp_1, \ldots, dp_n)$, the symplectic 2−form (Eq. 2.57) is written

$$\hat{\omega} = \sum_{i=1}^n dx^i \wedge dx^{n+i}. \tag{2.58}$$

We can compute symplectic k−forms by taking the k−fold *exterior products* of $\hat{\omega}$

$$\hat{\omega}_r = \sum_i dq^i \wedge dp_i,$$

$$\hat{\omega}_r \wedge \hat{\omega}_r = -2! \sum_{i_1 < i_2} dq^{i_1} \wedge dq^{i_2} \wedge dp_{i_1} \wedge dp_{i_2},$$

$$\hat{\omega}_r \wedge \hat{\omega}_r \wedge \hat{\omega}_r = -3! \sum_{i_1 < i_2 < i_3} dq^{i_1} \wedge dq^{i_2} \wedge dq^{i_3} \wedge dp_{i_1} \wedge dp_{i_2} \wedge dp_{i_3},$$

$$\cdots \cdots = \cdots \cdots \tag{2.59}$$

The largest $2n$−form is

$$\overbrace{\hat{\omega}_r \wedge \cdots \wedge \hat{\omega}_r}^{n-fold} = n!(-)^{[n/2]} dq^1 \wedge \cdots \wedge dq^n \wedge dp_1 \wedge \cdots \wedge p_n \tag{2.60}$$

and this defines the **oriented volume** form

$$\Omega_{\hat{\omega}} = \frac{(-)^{[n/2]}}{n!} \overbrace{\hat{\omega} \wedge \cdots \wedge \hat{\omega}}^{n-fold}. \qquad (2.61)$$

$[n/2]$ is the largest integer smaller than or equal to $n/2$.

As is illustrated in the Appendix A.9.2, the geometric meaning of forms is that of an area or volume, objects, which are quite often introduced in chemical theories. For example, reaction rates are determined by the flux through a multidimensional dividing surface (transition state) and the evaluation of the density of states of reactant molecules, both requiring the calculation of phase space areas and volumes [2].

Summarizing, $\hat{\omega}$ is a symplectic form on a manifold M of even dimension $2n$ and it is non-degenerate, skew-symmetric, closed $2-$form ($d\hat{\omega} = 0$). A pair $(M, \hat{\omega})$ is said to be a *symplectic manifold*. Those charts (coordinates) which satisfy Darboux's theorem, $\hat{\omega} = \sum_{i=1}^{n} dx^i \wedge dx^{n+i}$, are said to be *symplectic charts* and the local coordinates are called *canonical coordinates*. In the following we shall see that Hamiltonian mechanics and its geometrical   properties can be formulated with $\hat{\omega}$.

2.5.1 Symplectic Transformations

The equations

$$X_i = F_i(x, t), \quad i = 1, \ldots, 2n, \qquad (2.62)$$

define a transformation, which may involve both coordinates and their conjugate momenta, and do not change the equations of motion. These transformations are called *canonical* and the **Jacobian matrix**, (DF), of the transformation, $(DF)_{ij} = \partial F_i / \partial x_j$, satisfies the *symplectic property*

$$(DF)^T J (DF) = J. \qquad (2.63)$$

We can generalize the above transformations. A smooth map F that relates two symplectic manifolds $(M, \hat{\omega})$ and $(N, \hat{\tau})$ is said to be symplectic if $F^* \hat{\tau} = \hat{\omega}$, i.e., the pull-back of $\hat{\tau}$ yields $\hat{\omega}$. The symplectic maps are the canonical transformations of mechanics if the two manifolds (M, N) are identical

$$F^* \hat{\omega} = \hat{\omega}. \qquad (2.64)$$

Let $(M, \hat{\omega})$ be a symplectic manifold of dimension $2n$ with $\hat{\omega}$ a canonical symplectic $2-$form. The Hamiltonian function H is a smooth function on $M = T^*Q$. The Hamiltonian vector field, X_H, is then defined through the condition

$$i_{X_H} \hat{\omega} = \hat{\omega}(X_H, \bullet) = dH. \qquad (2.65)$$

$i_{X_H}\hat{\omega}$ symbolizes **interior product** and the triple $(M, \hat{\omega}, X_H)$ is a *Hamiltonian system*.

Indeed, we have seen that Hamilton's equations can be written in the form

$$\dot{x}(t) = J\partial H[x(t)] = X_{H_x}, \tag{2.66}$$

where H is the Hamiltonian function and X_{H_x} the Hamiltonian vector field at x. Since, the Hamiltonian vector field in local coordinates is written as

$$\left((X_H)^i, \overline{(X_H)}_j\right)^T = \left(\frac{\partial H}{\partial p_i}, -\frac{\partial H}{\partial q^j}\right)^T,$$

the 2−form $\hat{\omega}$ with a vacant position ($\bullet$) is transformed to

$$\begin{aligned}
\hat{\omega}(X_H, \bullet) &= \sum_{i=1}^{n}\left(dq^i(X_H)dp_i - dp_i(X_H)dq^i\right), \\
&= \sum_{i=1}^{n}\left((X_H)^i dp_i - \overline{(X_H)}_i dq^i\right), \\
&= \sum_{i=1}^{n}\left(\frac{\partial H}{\partial p_i}dp_i + \frac{\partial H}{\partial q^i}dq_i\right), \\
&= dH.
\end{aligned} \tag{2.67}$$

With every $Y \in \mathscr{X}(M)$ we can write

$$\hat{\omega}(X_H, Y) = dH(Y). \tag{2.68}$$

The integral curves of the Hamiltonian vector field X_H, $\Phi_t(x)$, are solutions of the canonical equations of motion Eq. 2.66. If the Hamiltonian does not have an explicit dependence on time, then, the energy is conserved. Indeed, the **Lie derivative** of the Hamiltonian is

$$\frac{d}{dt}H(x(t)) = dH(\dot{x}) = dH(X_{H_{x(t)}}) = \hat{\omega}(X_{H_{x(t)}}, X_{H_{x(t)}}) = 0. \tag{2.69}$$

We can also show this with charts.

$$\begin{aligned}
\hat{\omega}(X_{H_{x(t)}}, X_{H_{x(t)}}) &= \sum_i dq^i \wedge dp_i(X_{H_{x(t)}}, X_{H_{x(t)}}) \\
&= \sum_i \left[dq^i(X_{H_{x(t)}})dp_i(X_{H_{x(t)}}) - dp_i(X_{H_{x(t)}})dq^i(X_{H_{x(t)}})\right] \\
&= \sum_i \left[-\dot{q}^i \dot{p}_i + \dot{p}_i \dot{q}^i\right] = 0.
\end{aligned} \tag{2.70}$$

Symplectic diffeomorphisms $(F^*\hat{\tau} = \hat{\omega})$ leave Hamilton's equations invariant. Using the properties of pull-back (Eq. A.89) we show

$$i_{X_{F^*H_\tau}}\hat{\omega} = d(F^*H_\tau) = F^*dH_\tau = F^*i_{X_{H_\tau}}\hat{\tau} = i_{F_*^{-1}X_{H_\tau}}F^*\hat{\tau} = i_{F_*^{-1}X_{H_\tau}}\hat{\omega}, \quad (2.71)$$

which implies $X_{F^*H_\tau} = F_*^{-1}X_{H_\tau}$. $F_*^{-1}X_{H_\tau}$ means push-forward the vector field X_{H_τ} which is related with $\hat{\tau}$ to the vector field $X_{F^*H_\tau}$ associated with the symplectic 2–form $\hat{\omega}$.[4]

---

[4] For time dependent Hamiltonians, $H(q, p, t)$, we can apply the same formalism of conservative Hamiltonians by introducing time as a new variable, $q^0 = t$, with conjugate momentum, p_0, and new Hamiltonian, $H_t = p_0 + H(q, p, t) = 0$. Thus, the extended phase space $M_t = T^*Q_t$ of the *extended configuration manifold*, $Q_t = (t, q^1, \ldots, q^n)^T$, is of $2(n+1)$–dimension and in its cotangent bundle we define the Canonical Poincaré 1–Form

$$\hat{\theta}_t = \sum_{i=0}^{n} p_i dq^i = p_0 dq^0 + \sum_{i=1}^{n} p_i dq^i = -H(q, p, t)dt + \hat{\theta}, \quad (2.72)$$

and symplectic 2–form

$$\hat{\omega}_t = -d\hat{\theta}_t = dH \wedge dt - d\hat{\theta} = -dt \wedge dH + \sum_{i=1}^{n} dq^i \wedge dp_i. \quad (2.73)$$

The new Hamiltonian vector field (X_{H_t}) is defined by the equation

$$\hat{\omega}_t(X_{H_t}, \bullet) = dH_t, \quad (2.74)$$

$$\left(\frac{(X_{H_t})^0}{(X_{H_t})_0}\right) = \left(\begin{array}{c} 1 \\ -\partial H/\partial t \end{array}\right), \left(\frac{(X_{H_t})^i}{(X_{H_t})_i}\right) = \left(\begin{array}{c} \partial H/\partial p_i \\ -\partial H/\partial q^i \end{array}\right), \quad i = 1, \ldots, n. \quad (2.75)$$

The Hamiltonian vector field lives in the tangent bundle of the extended phase space, $T(T^*Q_t)$, the base vector fields of which are

$$\left(\frac{\partial}{\partial t}, \frac{\partial}{\partial p_0}\right), \left(\frac{\partial}{\partial q^i}, \frac{\partial}{\partial p_i}\right), \quad i = 1, \ldots, n. \quad (2.76)$$

$(M_t, \hat{\omega}_t, X_{H_t})$ is a Hamiltonian system and the Canonical Poincaré 1–Form, Eq. 2.72, is related to the total differential of action (Eq. 2.47). We can see, that with this formulation of time dependent systems the trajectories are projected at each time t in the physical phase space of the system of $2n$–dimension, $x = (q^1, \ldots, q^n, p_1, \ldots, p_n)^T$, and they are given by Hamilton's equations of motion with the time dependent Hamiltonian

$$\dot{x}(t) = J\partial H(x, t). \quad (2.77)$$

The symplectic maps of a symplectic vector space $(V, \hat{\sigma})$ onto itself

$$V : (V, \hat{\sigma}) \rightarrow (V, \hat{\sigma}), \quad F^*\hat{\sigma} = \hat{\sigma}, \tag{2.78}$$

form the *symplectic group Sp_{2n}*. Applying a symplectic transformation to the symplectic matrix J in local coordinate representation yields

$$(DF)^T J (DF) = J. \tag{2.79}$$

Theorem 1 (Liouville's Theorem)

If $(M, \hat{\omega}, X_H)$ is a Hamiltonian system, and Φ_t the flow of the vector field X_H $\left(\frac{d\Phi_t}{dt} = X_H\right)$, then, for all times t the flow is symplectic, i.e., $\Phi_t^* \hat{\omega} = \hat{\omega}$. From this, we conclude that the oriented volume $\Omega_{\hat{\omega}}$ (Eq. 2.61) is conserved (Liouville's Theorem).

2.5.2 Poisson Brackets

f and g are two dynamical quantities acting on the Hamiltonian system $(M, \hat{\omega}, H)$. If X_f and X_g are vector fields assigned to the two dynamical quantities, then, they are defined by the equations

$$\hat{\omega}(X_f, \bullet) = df, \quad \hat{\omega}(X_g, \bullet) = dg, \tag{2.80}$$

which imply

$$X_f = \left(\frac{\partial f}{\partial p}, -\frac{\partial f}{\partial q}\right)^T \quad \text{and} \quad X_g = \left(\frac{\partial g}{\partial p}, -\frac{\partial g}{\partial q}\right)^T. \tag{2.81}$$

The Poisson bracket is defined as

$$\hat{\omega}(X_f, X_g) = df(X_g)$$
$$= \sum_{i=1}^{n} \left[\frac{\partial f}{\partial q^i} dq^i (X_g) + \frac{\partial f}{\partial p_i} dp_i (X_g)\right]$$
$$= \sum_{i=1}^{n} \left[\frac{\partial f}{\partial q^i} \frac{\partial g}{\partial p_i} + \frac{\partial f}{\partial p_i} \left(-\frac{\partial g}{\partial q^i}\right)\right]$$
$$= \sum_{i=1}^{n} \left[\frac{\partial f}{\partial q^i} \frac{\partial g}{\partial p_i} - \frac{\partial f}{\partial p_i} \frac{\partial g}{\partial q^i}\right] \tag{2.82}$$
$$\equiv \{f, g\} = -\{g, f\}. \tag{2.83}$$

The Lie derivative of a dynamical quantity g with respect to a vector field X_f is defined as the directional derivative of g along the vector X_f

$$L_{X_f} g = dg(X_f) = \hat{\omega}(X_g, X_f). \qquad (2.84)$$

So, to be consistent with the definition of Poisson brackets (Eq. 2.82) for a Hamiltonian vector field we take $L_{X_H} g = dg(X_H) = \hat{\omega}(X_g, X_H) = \{g, H\}$.

Some properties of Poisson brackets are:

P1: The Poisson bracket in terms of Lie derivative is written as

$$\{g, f\} = L_{X_f} g = dg(X_f) = -df(X_g) = -L_{X_g} f = -\{f, g\}. \qquad (2.85)$$

P2: The quantity f (or g) is constant along the flow of X_g (X_f) if and only if $\{g, f\} = 0$.

P3: Let Φ_t be the flow of the Hamiltonian vector field X_H and g being a dynamical quantity, then, it is valid

$$\frac{d}{dt}(g \circ \Phi_t) = \frac{\partial}{\partial t}(g \circ \Phi_t) + \{g \circ \Phi_t, H\}. \qquad (2.86)$$

P4: Poisson brackets defined on the set of smooth functions $\mathscr{F}(M)$ on M generate a Lie algebra, i.e.,

- $\{f, g\}$ is bilinear,

- $\{f, f\} = 0$, and

- $\{f, \{g, h\}\} + \{g, \{h, f\}\} + \{h, \{f, g\}\}$ (Jacobi identity).

P5: In a local symplectic chart with canonical coordinates (q^i, p_j) the following equations are true

$$\{q^i, q^j\} = 0 \qquad (2.87)$$
$$\{p_i, p_j\} = 0 \qquad (2.88)$$
$$\{q^i, p_j\} = \delta^i_j. \qquad (2.89)$$

P6: F is diffeomorphism between two symplectic manifolds, $F : (M, \hat{\omega}) \to (N, \hat{\tau})$. This map is also symplectic if preserves the Poisson brackets of functions and/or $1-$forms, i.e.,

$$\{F^* f, F^* g\} = F^*\{f, g\} \text{ forall } f, g \in \mathscr{F}(N). \qquad (2.90)$$

Similarly to the previous section we can use the formalism of interior product to describe Lie derivatives and Poisson brackets. The Lie derivative of a form α is defined as (Cartan's magic formula)

$$L_X \alpha = i_X d\alpha + d i_X \alpha. \tag{2.91}$$

If α is a function (0−form) then

$$L_X \alpha = i_X d\alpha. \tag{2.92}$$

A differential form is conserved if

$$L_X \alpha = 0. \tag{2.93}$$

An example is the conservation of the canonical symplectic 2−form, $\hat{\omega}$, along a Hamiltonian vector field X_H

$$L_{X_H} \hat{\omega} = i_{X_H} d\hat{\omega} + d i_{X_H} \hat{\omega} = -i_{X_H} d \circ d\hat{\theta} + d \circ dH = 0. \tag{2.94}$$

The Poisson bracket is defined in terms of interior products as

$$\{g, f\} = L_{X_f} g = i_{X_f} dg = i_{X_f} i_{X_g} \hat{\omega}. \tag{2.95}$$

References

1. Arnold VI (1980) Mathematical methods of classical mechanics., Graduate text in mathematics Springer, Berlin
2. Ezra GS, Waalkens H, Wiggins S (2009) Microcanonical rates, gap times, and phase space dividing surfaces. J Chem Phys 130(164):118
3. Frankel T (2004) The geometry of physics: an introduction. Cambridge University Press, Cambridge
4. Ginoux JM (2009) Differential geometry applied to dynamical systems., Nonlinear science, World Scientific Publishing Co., Pte. Ltd, Singapore
5. Meyer KR, Hall GR, Offin D (2009) Introduction to Hamiltonian dynamical systems and the n-body problem, 2nd edn., Applied mathematical sciences Springer, Heidelberg
6. Scheck F (1990) Mechanics, 1st edn. Springer, Berlin
7. Spivak M (1965) Calculus on manifolds: a modern approach to classical theorems of advanced calculus. Addison-Wisley, Massachusetts

Chapter 3
Dynamical Systems

A general dynamical system is described by the system of ordinary differential equations

$$\frac{dx}{dt} \equiv \dot{x} = F(x, \mu), \tag{3.1}$$

where the independent variables are $x \in \mathbb{R}^n$ and $\mu \in \mathbb{R}^p$ are parameters of the system. The RHS of the equations, $F = (f^1(x, \mu), \ldots, f^n(x, \mu))^T$, are smooth functions and define a vector field or the velocity field $\dot{x}$.

Here, we are interested in Hamiltonian systems and flows in even dimensional phase space, $(M, \hat{\omega}, X_H)$. $(M, \hat{\omega})$ is a symplectic manifold of dimension $2n$ with $\hat{\omega}$ a canonical symplectic 2-form. The Hamiltonian function H is a smooth function on $M = T^*Q$ with Q signifying the configuration manifold and T^*Q the cotangent bundle of Q. The Hamiltonian vector field, X_H, is then defined with the relation

$$i_{X_H}\hat{\omega} = \hat{\omega}(X_H, \bullet) = dH. \tag{3.2}$$

$i_{X_H}\hat{\omega}$ indicates the interior product of the Hamiltonian vector field X_{H_x} with the 2-form $\hat{\omega}$. If $x = (q^1, q^2, \ldots, q^n, p_1, p_2, \ldots, p_n)^T$ defines a chart (coordinate system) in phase space M, then, the Hamiltonian vector field belongs to the tangent bundle of phase space $(TM \equiv T(T^*Q))$

$$\dot{x}(t) = J\partial H(x, \mu), \tag{3.3}$$

where ∂H is the gradient of Hamiltonian function, and J the symplectic matrix

$$J = \begin{pmatrix} 0_n & I_n \\ -I_n & 0_n \end{pmatrix}. \tag{3.4}$$

0_n and I_n are the zero and unit $n \times n$ matrices, respectively.

The study of dynamical systems is not simply restricted to distinguish quasiperiodic from chaotic flows (see below), although the onset of chaos by varying the

© The Author(s) 2014   33
S.C. Farantos, *Nonlinear Hamiltonian Mechanics Applied to Molecular Dynamics*,
SpringerBriefs in Electrical and Magnetic Properties of Atoms, Molecules, and Clusters,
DOI 10.1007/978-3-319-09988-0_3

Table 3.1 Low dimensional model potentials employed to study elementary bifurcations of equilibria and periodic orbits

Bifurcation	Potential
Center-Saddle (Saddle-Node)	$V(q) = \frac{1}{3}q^3 - \frac{1}{2}\alpha q^2 - \beta q - \gamma$
Pitchfork	$V(q) = \frac{1}{4}q^4 - \frac{1}{3}\alpha q^3 - \frac{1}{2}\beta q^2 - \gamma q - \delta$
Period doubling and $m : n$ resonances	$V(x, y) = \frac{1}{2}\left(\omega_x^2 x^2 + \omega_y^2 y^2\right) - \varepsilon x^2 y$
Complex unstable (Hamiltonian Hopf)	$V(x, y, z) = \frac{1}{2}\left(\omega_x^2 x^2 + \omega_y^2 y^2 + \omega_z^2 z^2\right) - \varepsilon x^2 y - \eta x^2 z$

parameters in nonlinear dynamical systems was a central point in numerical investigations during the first years of the development of the theory. The description and location of *time invariant structures* in phase space of nonlinear dynamical systems has been an exciting endeavour. By invariant phase space structures we mean equilibria, periodic orbits, tori, normally hyperbolic invariant manifolds and generally stable/unstable manifolds.

Trajectories initialized on these objects will remain on their surfaces for ever. Nearby trapped trajectories to these objects determine the dynamical behaviour of the system, and as we shall see, for molecules they can explain observed spectroscopic characteristics and reaction dynamics. Of great importance is the evolution of these invariant structures by varying parameters in the Hamiltonian or global constants of motion like the total energy. At critical values of the parameters *bifurcations (branching)* of these invariant structures are observed, which indicate the genesis of new qualitatively different motions of the system. One of the achievements of nonlinear mechanics is the classification of *elementary bifurcations*, which mean bifurcations described by very simple low dimensional Hamiltonians, nevertheless, equivalently encountered in generic multidimensional dynamical systems.

In this chapter we describe the different type of elementary bifurcations by numerically studying simple 1D, 2D and 3D model potentials tabulated in Table 3.1.

3.1 Equilibria and Elementary Bifurcations

Equilibria are defined as the solutions of the equations

$$\dot{x}(t) = 0. \tag{3.5}$$

For Hamiltonians written as the sum of kinetic energy, which depends only on momenta and potential energy, $V(q)$,

$$H(q, p) = \frac{1}{2}p^2 + V(q), \tag{3.6}$$

the equilibria are the critical points of the potential, i.e., $\partial V(q) = 0$. There are excellent books [9, 21] which describe the elementary bifurcations of fixed points of vector fields. Here, we do not attempt a complete cover of this subject but to discuss mainly those bifurcations which are met frequently with molecular potentials and particularly center-saddle bifurcations.

3.1.1 Cubic Potential

We assume a general cubic potential

$$V(q) = \frac{1}{3}q^3 - \frac{1}{2}\alpha q^2 - \beta q - \gamma. \tag{3.7}$$

The equilibrium points of Hamiltonian vector field are the roots of the second order polynomial

$$dV(q)/dq = q^2 - \alpha q - \beta = 0. \tag{3.8}$$

In order to Eq. 3.8 has two real roots (equilibrium points) the discriminant of the equation, $D = \alpha^2 + 4\beta$, should satisfy, $D \geq 0$. Thus, the parabola $D = 0$ defines the region in the two parameter space, (β, α), where these two roots exist. Figure 3.1a depicts this region and Fig. 3.1b shows the evolution of the two equilibria by varying the parameter β and for $\alpha = 0$. This graph is a typical *continuation/bifurcation diagram* (C/B). We notice, that there are no equilibrium points for negative values of β and at $\beta = 0$ the double root indicates the emanation of the *center-saddle* (CS) elementary bifurcation.[1] The two branches correspond to stable (solid line) and to unstable (dashed line) equilibria. Stable means that trajectories close to this point will remain in the nearby region, whereas unstable points mean that nearby trajectories will deviate from it. Visualization of the critical points of the corresponding potential function explains better the stable and unstable terms. Such a plot is shown in Fig. 3.1c for several values of β.

In this figure we distinguish three different regimes. For $\beta < 0$ there are no critical points, for $\beta > 0$ there are two critical points, one minimum and one maximum, and for $\beta = 0$ (dashed line) there is one critical point at $q = 0$. In Fig. 3.1d several trajectories are plotted in the phase plane, (q, p), and for $\beta = 1$ obtained by solving Eq. 3.3 . The dashed line depicts the *separatrix*, i.e., the line that separates the two distinctly different types of motion allowed for this dynamical system; closed stable orbits and unbound orbits. This phase space graph is typical of a center-saddle bifurcation and for integrable systems as 1D systems are. It is important to emphasize that the structure of phase space does not change qualitatively by introducing a second parameter. The C/B diagram remains the same with two branches for $\alpha \neq 0$. The

---

[1] For general dynamical systems this elementary bifurcation is called *saddle-node*.

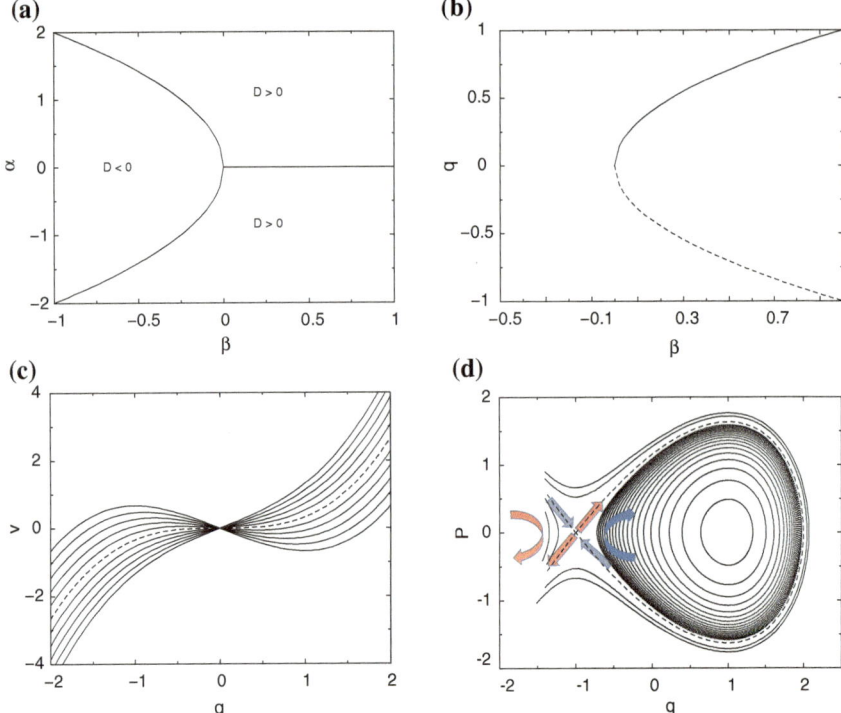

Fig. 3.1 Plots for a cubic potential. **a** The sign of the discriminant D in the parameter space (β, α) of a cubic potential. **b** Continuation/Bifurcation (C/B) diagram of a cubic potential. The coordinates q of the two equilibria are shown as function of the parameter β and for $\alpha = 0$. *Continuous line* indicates the stable equilibrium points (*minima*) and the *dashed line* the unstable equilibrium points (*maxima*). **c** The potential function for several values of β. For $\beta < 0$ there are no equilibria, for $\beta > 0$ there are two equilibrium points, one minimum and one maximum, and for $\beta = 0$ (*dashed line*) there is one saddle point at $q = 0$. **d** Trajectories which portray the phase space structure in the region of a center-saddle bifurcation ($\beta = 1$). The *dashed line* is the *separatrix*, an invariant curve which separates two different kinds of motion, bound and unbound. The *colored arrows* at the unstable equilibrium depict the direction of the linearized flow at the vicinity of equilibrium

colored arrows at the unstable equilibrium depict the direction of the linearized flow at the vicinity of equilibrium.

3.1.2 Quartic Potential

A general quartic potential is

$$V(q) = \frac{1}{4}q^4 - \frac{1}{3}\alpha q^3 - \frac{1}{2}\beta q^2 - \gamma q - \delta. \tag{3.9}$$

The critical points are the solutions of

$$dV(q)/dq = q^3 - \alpha q^2 - \beta q - \gamma = 0. \tag{3.10}$$

This cubic equation is reduced to a two parameter equation with the transformation

$$x = q - \alpha/3 \tag{3.11}$$

$$\mu = \frac{\alpha^3}{3} + \beta \tag{3.12}$$

$$\lambda = \frac{2\alpha^3}{27} + \frac{\alpha\beta}{3} + \gamma. \tag{3.13}$$

The reduced cubic polynomial is

$$x^3 - \mu x - \lambda = 0, \tag{3.14}$$

with a discriminant defined by

$$D = -\frac{\mu^3}{27} + \frac{\lambda}{4}. \tag{3.15}$$

The roots of Eq. 3.14 are classified according to:

1. For $D > 0$, there are one real root and two imaginary.
2. For $D < 0$, there are three different real roots.
3. $D = 0$, there are three real roots of which two of them are equal.

Figure 3.2a depicts the sign of the discriminant in the parameter space (λ, μ). The cusp curve defines the values of (λ, μ) where the discriminant is zero. Thus, crossing this curve from positive to negative values of D we pass from one to three equilibrium points. A double degeneracy of equilibrium points is encountered at the cusp curve. The three equilibria in the quartic potential are the two minima and one maximum of the potential. In Fig. 3.2b we plot potential curves for several values of λ. As λ approaches zero the kink in the potential is transformed to a double well.

The C/B diagram for $\lambda = 0$ and varying the parameter μ is shown in Fig. 3.3a. This is a typical *pitchfork* bifurcation. The introduction of a second parameter $(\lambda \neq 0)$ results in a C/B diagram shown in Fig. 3.3b. In other words, the unstable branch of the zero root becomes the unstable branch of a center-saddle bifurcation and one stable branch joins that of zero root. We can think of a continuation/bifurcation diagram as a folded surface in the (λ, μ, x) space.

Trajectories plotted in the phase plane are shown in Fig. 3.4 for the symmetric double well potential $(\lambda = 0$ and $\mu = 1)$. The *separatrix* (dashed line) emanated from the maximum of the potential separates the two types of motion encountered in this 1D system. The two types correspond to closed curves around a minimum and closed curves which encircle both minima. The colored arrows at the vicinity of the unstable equilibrium depict the linearized flow.

Fig. 3.2 a The sign of the
discriminant D (Eq. 3.15) in
the parameter space (λ, μ) of
a quartic potential. **b** Potential
curves of a quartic
polynomial and for several
values of λ. The *dashed line*
is the symmetric double well
potential $(\lambda = 0)$

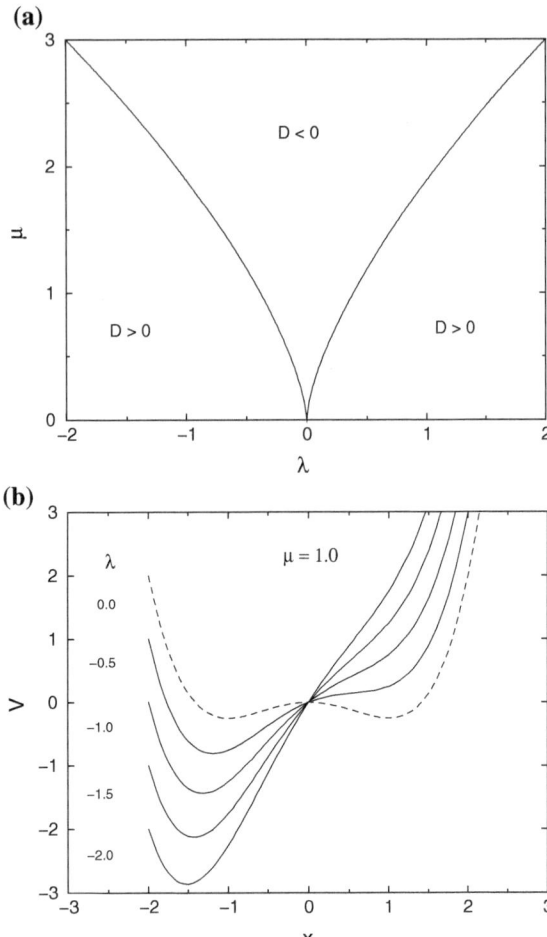

3.2 Periodic Orbits

$x(0)$ denotes the initial conditions of a trajectory at time $t_1 = 0$, then, this trajectory
is periodic if it returns to its initial point in phase space after the time $t_2 = T$
(period), i.e.,

$$x(T) - x(0) = 0. \tag{3.16}$$

 Thus, to find periodic solutions it is necessary to solve Eq. 3.3 subject to the *2-point
boundary conditions*, Eq. 3.16.

   The above boundary value problem is converted to an *initial value problem* by
considering the initial values of the coordinates and momenta s

$$x(0) = s, \tag{3.17}$$

Fig. 3.3
a Continuation/Bifurcation
diagram of a quartic
potential and for $\lambda = 0$
showing a *pitchfork*
bifurcation. **b** C/B diagram of
a quartic potential and for
$\lambda = 0.01$. A *center-saddle*
bifurcation and a continuous
branch emanate because of
the symmetry breaking for
$\lambda \neq 0$

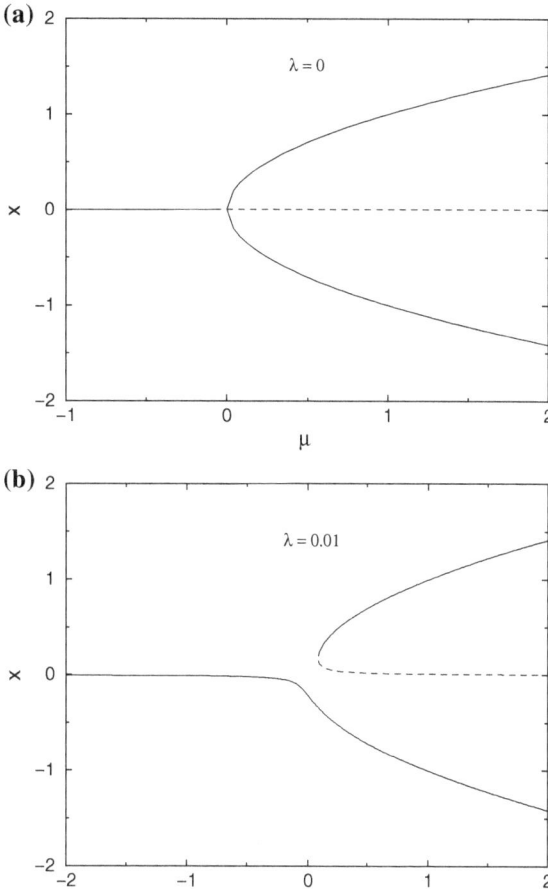

as independent variables in the nonlinear functions

$$B(s) = x(T) - s. \tag{3.18}$$

We symbolize the roots of Eq. 3.16 as s_*, i.e.,

$$B(s_*) = 0. \tag{3.19}$$

Thus, if s is a nearby value to the solution s_* we can compute the functions $B(s)$ by
integrating Hamilton's equations for the period T. By appropriately modifying the
initial values s we hope to converge to the solution, that is, $s \rightarrow s_*$ and $B \rightarrow 0$.
In Chap. 5 we describe the *multiple shooting method* for finding periodic orbits and
available software for carrying out such calculations.

Fig. 3.4 Phase space structure of a symmetric double well quartic potential. The *dashed line* is the *separatrix*, an invariant curve which separates two different kinds of motion. The *colored arrows* at the vicinity of the unstable equilibrium depict the linearized flow

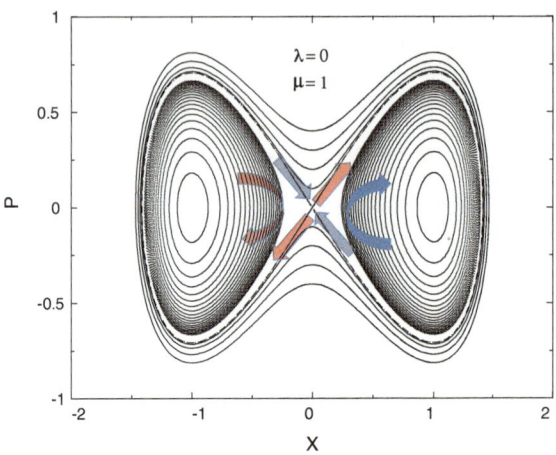

3.2.1 Stability Analysis of Equilibria and Periodic Orbits

To investigate the behaviour of neighbouring trajectories to an equilibrium by linearizing a vector field is a common strategy. The Grobman–Hartman theorem [11] states that the nonlinear flow is *locally topologically conjugate* to the flow of the linearized system in the vicinity of a hyperbolic equilibrium. Thus, important conclusions obtained for the linearized system can be extended to the nonlinear one.

If $x(t)$ is a solution of Hamilton's equation a nearby trajectory $x'(t)$ is given by the small displacement ζ

$$\delta x(t) = x'(t) - x(t) = \zeta(t), \tag{3.20}$$

the time evolution of which is described by

$$\dot{\zeta}(t) = \dot{x}'(t) - \dot{x}(t) = J\partial H(x') - J\partial H(x). \tag{3.21}$$

A Taylor expansion of the RHS of the above equation gives,

$$\dot{\zeta}(t) = J\partial^2 H[x(t)]\zeta(t) + h.o.t \ldots \tag{3.22}$$

$\partial^2 H[x(t)]$ denotes the matrix of second derivatives of the Hamiltonian (the Hessian) evaluated at $x(t)$. According to the discussion in Sect. A.7.4 of the Appendix, ζ is a *Jacobi field* and the Lie derivative along the Hamiltonian flow, $\Phi_{tx} = x(t)$, is zero. If, we define $A(t) = J\partial^2 H[x(t)]$, then, Eq. 3.22 is written,

$$\dot{\zeta}(t) = A(t)\zeta(t). \tag{3.23}$$

These are $2n$ linear differential equations with time dependent coefficients, and are called **variational equations**.

The general solution of variational equations is expressed as

$$\zeta(t) = Z(t)\zeta(0), \tag{3.24}$$

where $\zeta(0)$ is the initial displacement from a reference trajectory x and $Z(t)$ is the *fundamental matrix*, which also satisfies the variational equations as can be easily proved;

$$\dot{Z}(t) = A(t)Z(t). \tag{3.25}$$

If $x(t; s)$ denotes the reference trajectory with initial conditions s, then, we can show that the fundamental matrix has columns the vectors,

$$z^k = \frac{\partial x(t; s)}{\partial s^k}, \tag{3.26}$$

i.e., the derivatives of the trajectory $x(t; s)$ with respect to the initial coordinates s^k, $k = 1, \ldots, 2n$. Indeed, differentiating both sides of Eq. 3.3 we take,

$$\frac{\partial \dot{x}}{\partial s^k} = J \frac{\partial}{\partial s^k} \partial H,$$

$$\frac{d}{dt}\left(\frac{\partial x}{\partial s^k}\right) = J(\partial^2 H)\frac{\partial x}{\partial s^k}$$

$$= A(t)\frac{\partial x}{\partial s^k}. \tag{3.27}$$

Thus,

$$\dot{z}^k = A(t)z^k. \tag{3.28}$$

Obviously, at $t = 0$, Z is the matrix with columns the vectors $(1, 0, \ldots, 0)$, $(0, 1, 0, \ldots, 0), \ldots, (0, 0, \ldots, 1)$, i.e.,

$$Z(0) = I_{2n}. \tag{3.29}$$

For a periodic orbit with period T the fundamental matrix at $t = T$,

$$M = Z(T) = \frac{\partial x(T; s)}{\partial s}, \tag{3.30}$$

is named *monodromy matrix*. As we shall see, the monodromy matrix plays an important role in the theory of periodic orbits and their stability.

The behaviour of nearby trajectories to the periodic orbit, with the approximation of linearizing the equations of motion, is examined by studying the eigenvalues of the monodromy matrix λ_i. Here are some properties of the monodromy matrix.

1.
$$M(\dot{m}T) = M(T)M[(m-1)T]$$
$$M(mT) = M(T)^m \tag{3.31}$$

This is extracted from Eq. 3.24.

2. Since $\dot{x}$ is periodic in time as well as solution of the variational equations (Eq. 3.23), we can show that one of the eigenvalues of the monodromy matrix is equal to one. Indeed, for $\zeta(t) = \dot{x}(t)$, we have

$$\zeta(0) = \zeta(T) = M(T)\zeta(0)$$
$$(M - I_{2n})\zeta(0) = 0. \tag{3.32}$$

I_{2n} is the unit matrix. The above equation is true if at least one of the eigenvalues of the monodromy matrix is equal to one. Hence, if $x(t)$ is a periodic orbit, the variational equations have a periodic solution ($\dot{x}(t)$), and then, one of the eigenvalues of the monodromy matrix is one. The inverse is also true and very important; *if the monodromy matrix has one eigenvalue equal to one, then, the variational equations have a periodic solution.*

3. We can extend the previous property by proving that for every constant (integral) of motion there is one eigenvalue of the monodromy matrix equal to one. Let $C(x(0); T)$ a constant of motion along a periodic orbit with initial conditions $x(0) = s$ and period T. Then,

$$C(x(t; s)) = C(s). \tag{3.33}$$

Differentiating this equation with respect to the initial conditions

$$\partial_x C(x(t; s)) \frac{\partial x(t; s)}{\partial s} = \partial_s C(s), \tag{3.34}$$

which at $t = T$ becomes
$$\partial_s C(M - I_{2n}) = 0. \tag{3.35}$$

Thus, we conclude that the monodromy matrix has one eigenvalue equal to one, provided $\partial_s C \neq 0$.

4. For conservative Hamiltonians and Liouville's theorem we deduce the determinant of the monodromy matrix

$$det\, M(T) = 1. \tag{3.36}$$

5. The monodromy matrix satisfies the symplectic property

$$M^T J M = J. \tag{3.37}$$

From the above two properties we may conclude that the variational equations of a Hamiltonian system (the linearized system) is also Hamiltonian with the symplectic property (see also Eq. 2.74). This is already known, since the fundamental matrix is the Jacobian of the Hamiltonian flow, $\Phi_t(x_0)$, which defines an one-parameter group of symplectic diffeomorphisms, Sect. 2.5.1

$$Z(t) = D\Phi_t(x_0) \equiv \Phi_{t*}. \tag{3.38}$$

6. For a conservative Hamiltonian system if λ is an eigenvalue of the monodromy matrix obtained by solving the linearized approximation, then, λ^{-1} and their complex conjugates, λ^* and $(\lambda^*)^{-1}$ are also eigenvalues of M. In other words, for conservative Hamiltonian systems the eigenvalues of the monodromy matrix occur in pairs, (λ, λ^*) for complex eigenvalues and (λ, λ^{-1}) for real eigenvalues. According to the properties 2 and 3 for conservative Hamiltonian systems the monodromy matrix has always two eigenvalues equal to one.
From Eq. 3.31 we infer

$$\zeta(mT) = M(T)^m \zeta(0) = \begin{pmatrix} \lambda_1^m & 0 & ... & 0 \\ 0 & \lambda_2^m & ... & 0 \\ ... & ... & ... & ... \\ 0 & 0 & ... & \lambda_{2n}^m \end{pmatrix} \zeta(0), \tag{3.39}$$

assuming that the coordinate system is defined by the eigenvectors of M. Obviously, the eigenvalues of the monodromy matrix determine the deviation of a trajectory from the periodic orbit with an initial displacement $\zeta(0)$ after m iterations.

Sometimes it is convenient to express the eigenvalues of the monodromy matrix as,

$$\lambda = \exp(\alpha T), \tag{3.40}$$

the exponents α are called *characteristic exponents*.

From the properties of the monodromy matrix we may infer the following for its eigenvalues. Excluding the two unit eigenvalues:

(i) If all the eigenvalues are on the unit complex circle, and multiple eigenvalues (degenerate) have independent eigenvectors which are equal to the multiplicity of the eigenvalues, then, the periodic orbit is *stable (elliptic)*, and it is surrounded by tori. The characteristic exponents are pure imaginary numbers, $\alpha = i\sigma$, and σ may be considered as the frequency of rotation of a neighbouring trajectory around the periodic orbit. It may happen that,

$$T/(2\pi/\sigma) = m/n, \tag{3.41}$$

where m and n are integers. Then, it can be shown that the variational equations have a periodic solution, and there should exist a new periodic orbit of period $T' = nT$ in the neighborhood of the initial one.

Fig. 3.5 Eigenvalues of the monodromy matrix with respect to the complex unit circle. *Top* from left to right; The two pairs of eigenvalues are complex conjugate with norm equal to one (stable orbit), one pair of complex conjugate with unit norm and one pair of real eigenvalues (single unstable periodic orbit). *Bottom* from left to right; two real pairs (double unstable periodic orbit) and a quadruplet of complex eigenvalues out of unit circle (complex unstable periodic orbit)

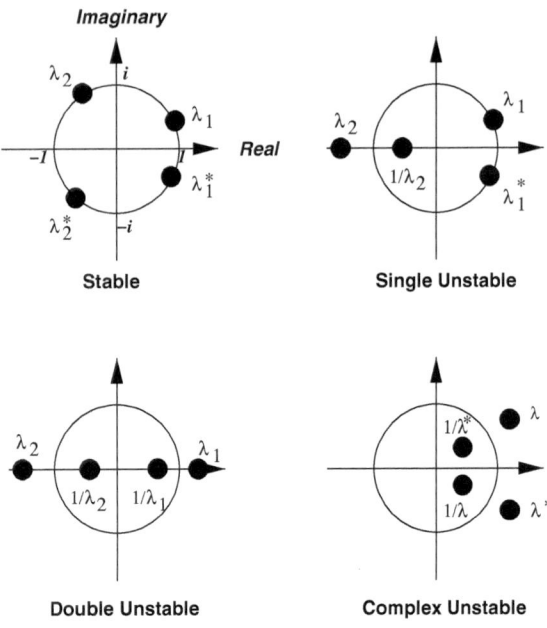

(ii) If there are eigenvalues equal to one, then, $\alpha = 0$ and a new periodic orbit of period T also exists in the neighborhood of the parent one. For eigenvalues equal to -1, $\alpha = i\pi/T$ and a new periodic orbit of double period exists nearby.

(iii) If there are real eigenvalues greater than ± 1, $\zeta(t)$ will deviate exponentially with time, and the periodic orbit is *unstable (hyperbolic)* in the directions of the corresponding eigenvectors.

(iv) If there is a complex eigenvalue, λ, with norm greater than one, then, λ^{-1}, λ^*, and $(\lambda^*)^{-1}$ are also eigenvalues, and the periodic orbit is called *complex unstable*.

(v) If there are multiple eigenvalues with independent eigenvectors less than the multiplicity of the eigenvalue, the periodic orbit is unstable with $\zeta(t)$ deviating not exponentially but with a power of t.

In Fig. 3.5 we plot the eigenvalues of M on the unit complex circle for a system of three degrees of freedom. The two pairs of eigenvalues which are different than one may be: complex conjugate with norm equal to one (stable orbits), or one pair of complex conjugate with unit norm and one pair of real eigenvalues (single unstable periodic orbits), or two real pairs for double unstable periodic orbits. There are three different cases in the double instability; positive, negative, or one positive and one negative pairs of real eigenvalues. Finally, if there is a quadruplet of complex eigenvalues out of the unit circle the periodic orbit is complex unstable.

The theories to predict when a stable periodic orbit will become unstable by varying a parameter of the Hamiltonian, as well as the number of branches at the bifurcation critical value of the continuation/bifurcation diagram are those of Krein,

Gelfand, and Lindskii [22] and degree theory [13]. In Chap. 5 we discuss numerical methods to solve 2-point boundary value problems.

3.2.2 Complex Instability

Three degrees of freedom systems exhibit new dynamical phenomena, such as the *Arnold diffusion* [9], i.e., the non isolation of the chaotic regions in phase space, and the phenomenon of *complex instability* [3]. The latter is related to unstable periodic orbits whose monodromy matrix has a quadruplet of complex eigenvalues out of the unit circle (Fig. 3.5). This can happen only for systems with three and more degrees of freedom. Around complex unstable periodic orbits the trajectories diverge exponentially while rotating with a characteristic frequency.

Specifically, the system which we have employed to investigate complex insta-bility is described by the Hamiltonian [3]

$$ H = \frac{1}{2}(p_x^2 + p_y^2 + p_z^2) + \frac{1}{2}(\omega_x^2 x^2 + \omega_y^2 y^2 + \omega_z^2 z^2) - \varepsilon x^2 y - \eta x^2 z. \quad (3.42) $$

The values of the parameters are, $\omega_x^2 = 0.9$, $\omega_y^2 = 1.6$, $\omega_z^2 = 0.4$, $\varepsilon = 0.08$, and $\eta = 0.01$. The harmonic frequencies satisfy the following resonance conditions; $\omega_x : \omega_y : \omega_z = 3:4:2$. This Hamiltonian supports a complex unstable family of periodic orbits for an extended energy domain. In Fig. 3.6 we depict the projection of the spiral invariant curve from a complex unstable periodic orbit at energy of 6 and initial conditions, $x = 0.0$, $p_x = 3.3711353$, $y = 0.6230596$, $p_y = 0.0$, $z = 0.1892212$, $p_z = 0.0$. The analytically obtained points (open squares) are compared with the numerically calculated points (filled squares), which in most of the cases, coincide (for details see [3]).

It turns out, that center-saddle bifurcations and complex unstable periodic orbits are frequently encountered in molecular dynamics, and indeed, at the beginning it was a surprise to chemists. Contrary to that, these phenomena are well known to mathematicians as the literature reveals. We quote from the book of Stephen Wiggins "Introduction to Applied Nonlinear Dynamical Systems and Chaos", p. 283 [21], ... *we might conclude that, in one-parameter families of vector fields, the most "typical" bifurcations are saddle-node and Poincaré–Adronov–Hopf.* Center-saddle bifurcations are typical in molecular Hamiltonian vector fields as well, and in Chap. 6 we discuss their striking impact to molecular dynamics with experimental manifestations.

3.2.3 Existence Theorems for Periodic Orbits

Knowing the existence of PO is essential in the process of locating them. As a matter of fact, numerically locating periodic orbits in Hamiltonian systems and continuing

Fig. 3.6 Projection of the spiral invariant curve from a complex unstable periodic orbit. The analytically obtained points (*open squares*) are compared with the numerically calculated points (*filled squares*), which in most of the cases, coincide [3]

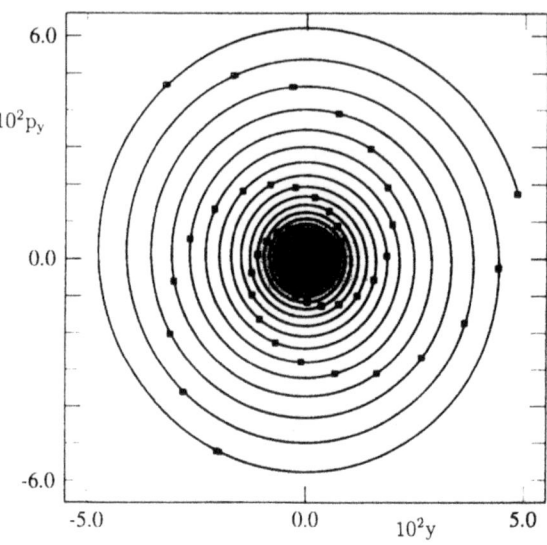

them in the parameter space is as science as 'art'. In any case, finding a family of periodic obits one needs the 'seed' to start the process of locating the PO as well as the continuation. Hence, existence theorems are vital.

Poncaré–Birkhoff theorem [2, 10] guarantees the appearance of alternating PO, stable–unstable, by perturbing commensurate tori in conservative systems. Another important existence theorem of periodic orbits is that of Weinstein [20]. This theorem guarantees, that arbitrarily close to a stable equilibrium point of a n degrees of freedom system there are at least n periodic orbits whose periods are close to those of the linearized system. Generalizations of this theorem to unstable equilibria were given by Moser [15]. A survey of the existence theorems of periodic orbits for nonlinear dynamical systems may be found in [17, 23].

Center-saddle bifurcations of periodic orbits are ubiquitous and their existence is supported by the Newhouse theorem [16, 21] initially proved for dissipative dynamical systems, and it was later extended to Hamiltonian systems as well [4, 8]. The theorem states that tangencies of the stable and unstable manifolds associated with unstable equilibria and periodic orbits, generate an infinite number of period doubling and center-saddle bifurcations. Hence, the originally single unstable periodic orbits (named Lyapunov orbits) of the fundamental families of index-1 equilibria are expected to generate such CS bifurcations as their manifolds expand along the unstable degree of freedom (Sect. 3.4). As a matter of fact, intersections of the stable and unstable manifolds coming either from the same or different equilibria generate *homoclinic* and *heteroclinic* orbits, respectively [21], which connect remote regions of phase space. Their numerical location is not easy and that makes periodic orbit families more precious in studying the complexity of the molecular phase space at high excitation energies.

3.3 Tori and Normally Hyperbolic Invariant Manifolds

A well known integrable system to chemists is the vibrational Hamiltonian of a molecule expressed as sum of harmonic oscillators. To show this we expand the global Hamiltonian in a Taylor series around an elliptic equilibrium x_0, which is taken to be the origin of the coordinate system. Then, assuming the constant term to be zero the Taylor series is written as

$$H(x) = \sum_{l=0}^{\infty} \frac{1}{l!} H_l(x). \tag{3.43}$$

H_l is a homogeneous polynomial of degree $l + 2$ in the variables $x = (q^1, \ldots, q^n, p_1, \ldots, p_n)^T$.

$H_0(x)$ denotes the quadratic part of the Hamiltonian, $H_0(x) = \frac{1}{2} x^T [\partial^2 H(0)]x$, where $[\partial^2 H]$ is the Hessian of the Hamiltonian evaluated at the equilibrium point. The linearized Hamiltonian vector field is defined by

$$\dot{x}(t) = J[\partial^2 H(0)]x(t) = Ax(t). \tag{3.44}$$

By diagonalizing the matrix, $A = J[\partial^2 H(0)]$, we obtain the normal coordinates of the molecule which render the quadratic Hamiltonian into a harmonic one[2]

$$H_0 = \frac{1}{2} \sum_{i=1}^{n} (p_i^2 + \omega_i^2 q^{i2}), \tag{3.45}$$

where ω_i are the frequencies of the normal modes.

In the Appendix A.10 we show that a quadratic Hamiltonian can be written in *action-angle* variables, (I_i, ϕ_i),

$$H_0 = \sum_{i=1}^{n} \omega_i I_i. \tag{3.46}$$

It is easy to show by writing Hamilton's equations that I_i are constants of motion and $\phi_i(t) = \omega_i t + \phi_{i0}$, where ϕ_{i0} are the initial phases. Thus, the flow lies on a torus. An example of such a torus was presented in the introduction for the 2D model of carbon dioxide (Fig. 1.2).

A n-torus is the product of n circles, $T^n = \overbrace{S^1 \times S^1 \times \cdots \times S^1}^{n}$. This is a n-dimensional compact manifold embedded in a $2n$ − dimensional phase space. Trajectories with *commensurable* frequencies, i.e., $\sum_{i=1}^{n} m_i \omega_i = 0$ with m_i integer

---

[2] To avoid using many symbols we use (q, p) to ascribe both the internal and the normal coordinates and conjugate momenta.

numbers ($m_i \in \mathbb{Z}$) are periodic closed orbits, whereas for *incommensurable* frequencies, $\sum_{i=1}^{n} m_i \omega_i \neq 0$, the trajectories are *quasiperiodic* and cover the surface of n-torus.

3.3.1 Kolmogorov–Arnold–Moser Theorem

The very important theorem of Kolmogorov–Arnold–Moser guarantees that tori will survive even when we accept higher order terms in the Taylor expansion of the Hamiltonian, Eq. 3.43.

Theorem 3.1 (Kolmogorov-Arnold-Moser (KAM) Theorem)

A Hamiltonian system is written by $H = H_0 + \varepsilon H_1$, with H_0 an integrable Hamiltonian and εH_1 a small perturbation which makes H non-integrable. The KAM theorem guarantees that incommensurate tori, which exist for $\varepsilon = 0$, will also exist for $0 < \varepsilon \ll 1$. As ε increases the tori are destroyed and the trajectories become chaotic with no constants of motion.

3.3.2 Normally Hyperbolic Invariant Manifold

In the case of a hyperbolic equilibrium point instead of an elliptic one, for example a saddle of index-1, the quadratic part in the Taylor expansion of the Hamiltonian, Eq. 3.43, is written in scaled normal coordinates as

$$H_0 = \lambda Q^1 P_1 + \frac{1}{2} \sum_{i=2}^{n} \omega_i (P_i^2 + Q^{i2}), \tag{3.47}$$

where we have taken (Q^1, P_1) to be the unstable degree of freedom in the saddle. To go beyond the quadratic approximation by including higher order terms in the Taylor expansion of the Hamiltonian, one can apply the Poncaré - Birkhoff method to obtain the appropriate *normal form* coordinates (F, P_F) as are described in the Appendix A.10. We can go on by defining action-angle variables related to the normal form coordinates with the same transformations as applied to normal coordinates,

$$I_i = (F^{i2} + P_{Fi}^2)/2, \ i = 2, \ldots, n, \tag{3.48}$$

and for the unstable degree of freedom

$$J_1 = F^1 P_{F1}. \tag{3.49}$$

The local normal form Hamiltonian is written in terms of action variables only, $H_L(J_1, I_2, \ldots, I_n)$. Since, the transformation to the normal form coordinates is symplectic, Hamilton's equations do not change (see Eq. 2.71), and thus, one can write

$$
\begin{aligned}
\dot{F}^1 &= \frac{\partial H_L}{\partial P_{F^1}} = \frac{\partial H_L}{\partial J_1} \frac{\partial J_1}{\partial P_{F^1}} = \lambda_1(J_1, I_2, \ldots, I_n)\, F^1, \\
\dot{F}^i &= \frac{\partial H_L}{\partial P_{F^i}} = \frac{\partial H_L}{\partial I_i} \frac{\partial I_i}{\partial P_{F^i}} = \omega_i(J_1, I_2, \ldots, I_n)\, P_{F^i}, \quad i = 2, \ldots, n, \\
\dot{P}_{F^1} &= -\frac{\partial H_L}{\partial F^1} = -\frac{\partial H_L}{\partial J_1} \frac{\partial J_1}{\partial F^1} = -\lambda_1(J_1, I_2, \ldots, I_n)\, P_{F^1}, \\
\dot{P}_{F^i} &= -\frac{\partial H_L}{\partial P_{F^i}} = -\frac{\partial H_L}{\partial I_i} \frac{\partial I_i}{\partial F^i} = -\omega_i(J_1, I_2, \ldots, I_n)\, F^i, \quad i = 2, \ldots, n. \quad (3.50)
\end{aligned}
$$

λ_1 is the rate with which neighbouring trajectories diverge *transversely* along the unstable degree of freedom and ω_i, $i = 2, \ldots, n$, the vibrational frequencies in the stable degrees of freedom.

These equations can be used to define the normally hyperbolic invariable manifold (NHIM). Keeping $F^1 = P_{F^1} = 0$ we obtain the NHIM and for $F^1 = 0$ and $P_{F^1} \neq 0$ or $F^1 \neq 0$ and $P_{F^1} = 0$ we construct the stable/unstable manifolds, (W^s / W^u), of the NHIM.[3] For a $(2n)$D phase space the energy hypersurface is of $(2n - 1)$-dimension and a NHIM associated to a saddle index-1 is a sphere (S^{2n-3}) of $(2n - 3)$D in the energy shell. The stable/unstable manifolds of the NHIM, (W^s / W^u), are $(2n - 2)$D objects with the geometry that of a cylinder, $S^{2n-3} \times \mathbb{R}$ with the NHIM as equator.

NHIM with their stable/unstable manifolds have the correct dimension to act as dividing surfaces (transition states) in chemical reactions as Wiggins and collaborators have demonstrated in the last years [19].

3.4 Poincaré Surfaces of Section

Having the picture of a 2D torus embedded in a 4D phase space we take a section of this surface with a plane. Then, the quasiperiodic trajectory which covers the surface of the torus will cut the plane of section several times and it will leave traces that form a closed orbit (see Fig. 1.2). This is what we call Poincaré surface of section (PSS). We expect integrable systems to have PSS filled with closed regular curves. On the other hand, if the trajectory covers in infinite time the total energy surface, as chaotic ones do, we expect the trajectory to have points scattered in the plane of section, since the plane intersects the 3D energy surface in the 4D phase space.

As an example Fig. 3.7A shows a typical PSS for the model potential [5]

$$
V(x, y) = \frac{1}{2}\left(\omega_x^2 x^2 + \omega_y^2 y^2\right) - \varepsilon x^2 y. \quad (3.51)
$$

---

[3] Examples of stable/unstable manifolds (separatrices) are shown in Figs. 3.1d and 3.4, where the arrows explain the directions of the linearized flow in the neighbourhood of the unstable equilibrium.

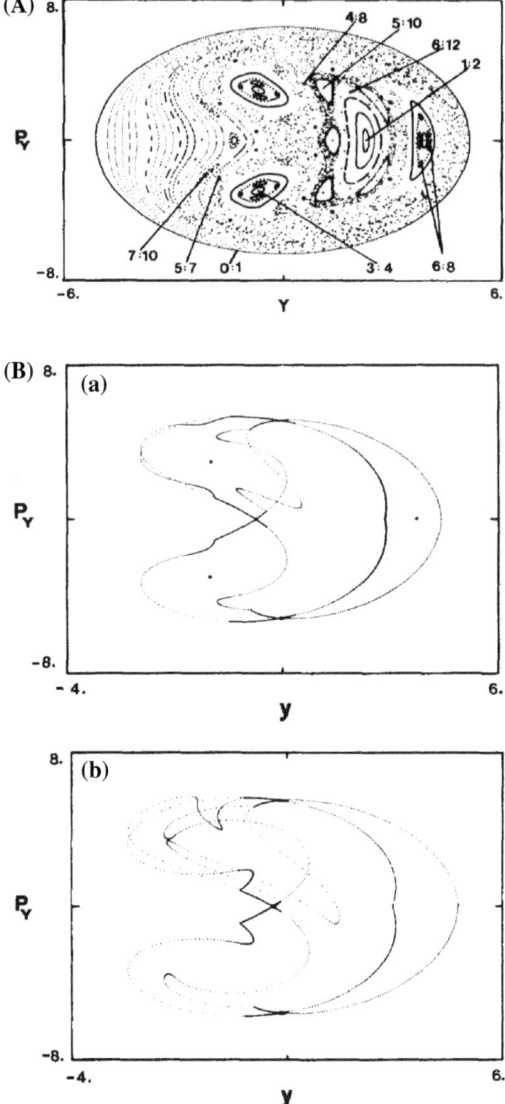

Fig. 3.7 **A** Barbanis–Contopoulos potential [5]: $V(x, y) = \frac{1}{2}\left(\omega_x^2 x^2 + \omega_y^2 y^2\right) - \varepsilon x^2 y$. The parameters used are $\omega_x^2 = 0.9$, $\omega_y^2 = 1.6$, $\varepsilon = 0.08$ and $\omega_x/\omega_y = 3/4$. The Poincaré surfaces of section for several trajectories on the (y, p) plane for $x = 0$ and $p_x > 0$ are shown. The intersections of several stable periodic orbits are also plotted and labelled. The scattered points among the islands of the 3:4 resonance are the intersections of one chaotic trajectory. Points near the border of the graph, 0/1, do not correspond to the chaotic trajectory but to several regular orbits. All trajectories correspond to total energy of about 21 units. **B** Inner and outer separatricies of the 3/4 resonance at energies (**a**) 23 and (**b**) 25. The double turnstiles can be seen in one of the islands which surround the elliptic points

The plane of section, (y, p), is defined by $x = 0$ and $p_x > 0$. The periodic orbits provide the 'skeleton' of phase space structure. In Fig. 3.7A the PSS of a few tori around the stable periodic orbits are shown at the energy of 21 units. The randomly scattered dots among the islands at 3/4, 5/7, and 5/10 resonances are the intersections of one chaotic trajectory with initial conditions close to the unstable 3/4 periodic orbit. The chaotic region increases with the energy. As the energy increases the branches which emerge from the 1/2 and 0/1 families go away from the parent family and proceed to enter the unstable region of 3/4 PO.

The inner and outer separatrices of the 3/4 resonance are shown in Fig. 3.7B for two different energies (a) $E = 23$ and (b) $E = 25$ units. These separatrices divide the phase plane in in- and out-resonance regions. To compute the separatrices two successive hyperbolic points are first chosen on the $x = 0$ plane of section. The eigenvectors of the monodromy matrix in the variational equations define the stable and the unstable manifolds associated with the hyperbolic points at the linear approximation. By taking initial conditions for a set of trajectories along these eigenvectors and propagating them forward in time for the unstable manifold and backward in time for the stable manifold for several periods, we construct numerically the separatrices. Each manifold can never intersect itself and that results in a complex pattern as the other hyperbolic point is approached. However, the unstable and stable manifolds can intersect and that leads to the formation of the turnstiles. There is an infinite number of intersections which define infinite homoclinic (heteroclinic) points. The homoclinic points correspond to trajectories which approach the hyperbolic points in infinite time from both forward and backward directions. According to MacKay et al. [14] and Bensimon and Kadanoff [1] theory the area of the turnstile gives the flux to get in (out) the resonance region and it can be calculated from the actions of homoclinic orbits.

To calculate PSS for a Hamiltonian system we usually take the plane, which the trajectory intersects, to be defined by one of the coordinates, for example $x^1 = c$, and requiring its conjugate momentum to be positive, $p_1 > 0$. The best method to compute PSS is that of Hénon [12], who suggested to replace the integration variable in equations of motion by the coordinate, say x^1. In other words, we replace the equations of motion

$$\frac{dx^i}{dt} = f^i(x), \ i = 1, \ldots, 2n, \tag{3.52}$$

by the equations

$$\frac{dt}{dx^1} = \frac{1}{f^1}, \ldots, \frac{dx^{2n}}{dx^1} = \frac{f^{2n}}{f^1}. \tag{3.53}$$

In practice, the trajectory is integrated in time and we check regularly when it crosses the plane of section. Once this has happened, the equations of motion are replaced by the Eq. 3.53 and we integrate them from the current point of x^1 to c. The x^1-axis need not be perpendicular to the Poincaré surface of section; any x^i can be chosen as integration variable provided the x^i-axis is not parallel to the Poincaré surface of section at the trajectory intersection point.

3.5 Non-periodic Trajectories

3.5.1 Maximal Lyapunov Exponent

Chaotic systems show a sensitivity to the initial conditions of trajectories. Tiny perturbations to the initial conditions lead to exponentially divergent neighbouring trajectories. A measure for the rate of divergence is the *Maximal Lyapunov Exponent* (MLE) [18]. MLE is obtained by integrating two initially close trajectories and estimating the rate of divergence of one trajectory with respect to the reference trajectory. Although, this quantity is defined for infinite integration time of the reference trajectory, in practice we integrate the trajectory initialized at the point $x(0)$ for finite times. This is a local property and it is computed by the formula

$$\Lambda = \lim_{\tau \to \infty} \frac{1}{\tau} \ln \left(\frac{||\delta x(\tau)||}{||\delta x(0)||} \right), \tag{3.54}$$

where $||\delta x(\tau)||$ usually means the Euclidean distance of the two trajectories at time τ. To avoid overflows in the exponential separation of the two trajectories, it is better to integrate for small time intervals and rescale the separation $\delta x(\tau_i)$ by the factor $\delta x(0)/\delta x(\tau_i)$. Then, we compute the sum

$$\Lambda = \lim_{\tau \to \infty} \frac{1}{\tau} \sum_i \ln \left(\frac{||\delta x(\tau_i)||}{||\delta x(0)||} \right), \quad \tau = \sum_i \tau_i. \tag{3.55}$$

Averaging over several neighbouring trajectories provide average local MLE, $< \Lambda > = \frac{1}{N} \sum_{i=1}^{N} \Lambda_i$.

It is not difficult to realize that by calculating the rate of divergence of nearby trajectories to the phase space point $x(0)$ and for infinitesimal displacements the quantity involved is nothing else than the Jacobi field (Eq. 3.22). In other words, what we compute is $\delta x(\tau) = Z(x(0), \tau)\delta x(0)$ with $Z(x(0), \tau)$ the fundamental matrix, Eq. 3.24. The maximal Lyapunov exponent is then expressed by the equation

$$\Lambda = \lim_{\tau \to \infty} \frac{1}{\tau} \ln \left(\frac{||Z(x(0), \tau)\delta x(0)||}{||\delta x(0)||} \right)$$

$$\approx \frac{1}{\tau} \ln \left(|\lambda_m(x(0), \tau)| \right), \tag{3.56}$$

where $\lambda_m(x(0), \tau)$ is the maximal eigenvalue of matrix Z at time τ.

3.5.2 Autocorrelation Functions

The average behaviour of a batch of trajectories in time can be traced by calculating autocorrelation functions (or survival probability functions). These are defined as follows

$$\Omega(t) = \int \rho[x(0)]\rho[x(t)]dx. \tag{3.57}$$

The classical initial distribution $\rho[x(0)]$ is very often assumed to be a Gaussian function centered at a chosen region of phase space. The spectrum is then defined as the Fourier transform,

$$I_c(\omega) = \int e^{i\omega t}\Omega(t)dt. \tag{3.58}$$

If we represent a conserved in time phase space distribution function, ρ, with a dynamical variable which depends on time, $g \circ \Phi_t = \rho[q(t), p(t), t]$, then, its Lie derivative along the Hamiltonian flow, Φ_t, is zero and this gives Liouville's equation (Sect. 2.5.2 (P3))

$$L_{X_H}g = \frac{d}{dt}(g \circ \Phi_t) = \frac{\partial}{\partial t}(g \circ \Phi_t) + \{g \circ \Phi_t, H\} = 0, \tag{3.59}$$

where $\{,\}$ is the Poisson bracket and H is the classical Hamiltonian. Liouville's equation is simply written as

$$\frac{\partial \rho}{\partial t} = -\{\rho, H\}. \tag{3.60}$$

The classical survival probability function, Eq. 3.57, and its Fourier transform, Eq. 3.58, have successfully been used in molecular dynamics [6, 7]. An example is shown in Fig. 3.8 for the 3D model potential tabulated in Table 3.1 [3].

Fig. 3.8 Comparison of the square amplitude of the quantum mechanical autocorrelation function (*continuous line*) with its classical mechanical analog (*broken line*) for the 3D model Hamiltonian given in Table 3.1 [3]

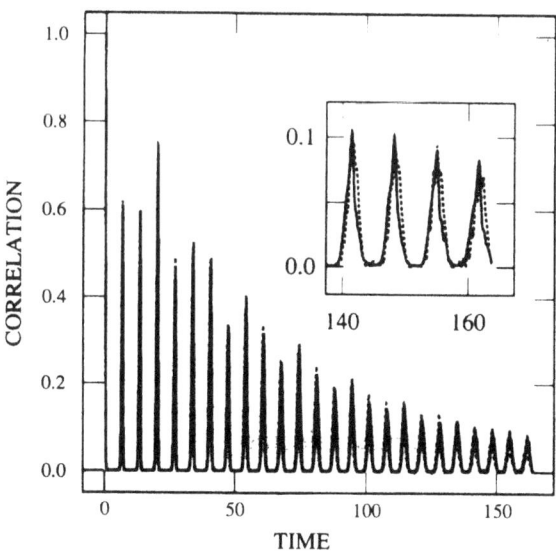

In the next chapter we examine how the classical survival probability function is related to its quantum mechanical analogue.

References

1. Bensimon D, Kadanoff LP (1984) Extended chaos and disappearance of KAM trajectories. Physica D 13:82–89
2. Birkhoff GD, Lewis DC (1933) On the periodic motions near a given periodic motion of a dynamical system. Ann Mat Pura Appl 12:117–133
3. Contopoulos G, Farantos SC, Papadaki H, Polymilis C (1994) Complex unstable periodic orbits and their manifestation in classical and quantum dynamics. Phys Rev E 50(5):4399–4403
4. Duarte P (1999) Abundance of elliptic isles at conservative bifurcations. Dyn Stab Sys 14(4):339–356
5. Founargiotakis M, Farantos SC, Contopoulos G, Polymilis C (1989) Periodic orbits, bifurcations and quantum mechanical eigenfunctions and spectra. J Chem Phys 91(1):1389–1402
6. Gomez Llorente JM, Taylor HS (1989) Spectra in the chaotic region: a classical analysis for the sodium trimer. J Chem Phys 91:953–962
7. Gomez Llorente JM, Pollak E (1992) Classical dynamics methods for high energy vibrational spectroscopy. Ann Rev Phys Chem 43:91–126
8. Gonchenko SV, Silnikov LP (2000) On two-dimensional area-preserving diffeomorphisms with infinitely many elliptic islands. J Stat Phys 101(1/2):321–356
9. Guckenheimer J, Holmes P (1983) Nonlinear oscillations, dynamical systems, and bifurcations of vector fields. Springer, Berlin
10. Hanssmann H (2007) Local and semi-local bifurcations in Hamiltonian dynamical systems: results and examples. Springer, Berlin
11. Hartman P (1964) Ordinary differential equations. Wiley, New York
12. Hénon M (1982) On the numerical computation of Poincaré maps. Physica D 5:412–414
13. Krasnosel'skii MA, Zabreiko PP (1984) Geometrical methods of nonlinear analysis. A series of comprehensive mathematics, Springer, Berlin
14. Mackay RS, Meiss JD, Percival IC (1984) Transport in Hamiltonian systems. Physica D 13:55–81
15. Moser J (1976) Periodic orbits near an equilibrium and a theorem by Alan Weinstein. Commun Pure Appl Math 29:727–747
16. Newhouse SE (1979) The abundance of wild hyperbolic sets and non-smooth stable sets for diffeomorphisms. Publ Math IHES 50:101–151
17. Rabinowitz PH (1984) Periodic solutions of Hamiltonian systems: a survey. SIAM J Math Anal 13:343
18. Skokos C (2010) The Lyapunov characteristic exponents and their computation. Lec Notes Phys 790:63–135
19. Waalkens H, Burbanks A, Wiggins S (2004) Phase space conduits for reaction in multidimensional systems: HCN isomerization in three dimensions. J Chem Phys 121(13):6207–6225
20. Weinstein A (1973) Normal modes for nonlinear Hamiltonian systems. Inv Math 20:47–57
21. Wiggins S (2003) Introduction to applied nonlinear dynamical systems and chaos, 2nd edn. Springer, New York
22. Yakubovich VA, Starzhinskii VM (1975) Linear differential equations with periodic coefficients. Wiley, New York
23. Yoshizawa T (1975) Stability theory and the existence of periodic solutions., Applied mathematical sciences, Springer, Berlin

Chapter 4
Quantum and Semiclassical Molecular Dynamics

Exploration of the structure of phase space (location of the phase space invariants) provides a detailed description of molecular dynamics. However, molecules are quantum objects and the study of quantum molecular dynamics should always be followed. In eighties, a plethora of numerical investigations of classical and quantum dynamics of molecules with model and realistic potential energy surfaces revealed that the eigenfunctions are localized in the same regions of phase space where invariant structures are. The importance of periodic orbits to identify these regions was emerged by demonstrating that eigenfunctions are scarred by unstable periodic orbits or the resonance region of stable periodic orbits [18, 19]. Even examples where eigenfunctions are influenced by the stable and unstable manifolds of unstable periodic orbits have been reported [31]. A clear distinction of the statistics of the eigenenergy distributions in cases of regular and chaotic classical dynamics was also identified [4].

In this chapter, after an introduction of the basic principles of quantum mechanics in their two main formulations, the canonical and the path integral, we present the theory for numerically solving Schrödinger equation for polyatomic molecules. The semiclassical analogue of the quantum mechanical propagator is also discussed.

4.1 Canonical Quantum Mechanics

To pass from classical to quantum mechanics we replace the $2n$-dimensional tangent bundle (TQ) of the configuration manifold Q of n-dimension with the *Hilbert space,* $\mathscr{H}$, of complex functions $(|\psi\rangle)$ that form a vector space of finite or infinite dimension. We use Dirac's notation to denote the vectors of Hilbert space, the *kets*, $|\psi\rangle$. Similarly to classical mechanics, we also work with the *dual* vector space of Hilbert space, $\mathscr{H}^*$, the set of linear transformations acting on $\mathscr{H}$ the vectors of which are denoted by the *bras*, $\langle\psi|$.

As both $\mathscr{H}$ and $\mathscr{H}^*$ are linear vector spaces, we can define a basis set $|e_i\rangle$ in $\mathscr{H}$ and its dual basis $\langle e^i|$ in $\mathscr{H}^*$, such that

© The Author(s) 2014

S.C. Farantos, *Nonlinear Hamiltonian Mechanics Applied to Molecular Dynamics,*
SpringerBriefs in Electrical and Magnetic Properties of Atoms, Molecules, and Clusters,
DOI 10.1007/978-3-319-09988-0_4

$$|\psi\rangle = \sum_j \psi^j |e_j\rangle$$

$$\langle\psi| = \sum_i \langle e^i|\psi_i = \sum_i \langle e^i|\psi^{i*}, \tag{4.1}$$

where the coefficients ψ^{i*1} are the complex conjugate of the coefficients ψ^i of the ket-vectors. Having introduced the dual of Hilbert space we can also define an inner product as a linear function that belongs to the dual Hilbert space

$$(|\phi\rangle, \bullet)(|\psi\rangle) \equiv \langle\phi|\psi\rangle = \sum_i \phi^{i*} \sum_j \psi^j \langle e^i|e_j\rangle$$

$$= \sum_i \phi^{i*} \psi^i. \tag{4.2}$$

The latter equation is the result of adopting an *orthonormal* basis set, $\langle e^i|e_j\rangle = \delta^i_j$.
With an orthonormal basis a state is represented as

$$|\psi\rangle = \sum_{k=1}^n c_k |e_k\rangle, \tag{4.3}$$

and the coefficients c_k are extracted by projecting $|\psi\rangle$ on the basis vector $|e_k\rangle$, $c_k = \langle e_k|\psi\rangle$. Thus,

$$|\psi\rangle = \sum_{k=1}^n \langle e_k|\psi\rangle|e_k\rangle = \sum_{k=1}^n |e_k\rangle\langle e_k|\psi\rangle. \tag{4.4}$$

The above equation implies the *completeness relation*

$$\sum_{k=1}^n |e_k\rangle\langle e_k| = I_n. \tag{4.5}$$

I_n is the identity operator in $\mathscr{H}$, a unit $n \times n$ matrix.
Canonical quantization is introduced by the following axioms:

1. For an *isolated* system the state vectors $|\psi\rangle$ form a Hilbert complex vector space.
2. For every classical observable, i.e., a function in phase space $O(q, p)$, there is a *Hermitian operator*,[2] $\hat{O}$, acting on $\mathscr{H}$.

---

[1] We have defined $\psi_i = \psi^{i*}$.

[2] Hermitian is the operator which is equal to its adjoint, $\hat{O} = \hat{O}^\dagger$. Having defined a basis set the representation of the Hermitian operator satisfies $\langle e^i|\hat{O}|e_j\rangle = \langle e^i|O^\dagger|e_j\rangle = (\langle e^j|\hat{O}|e_i\rangle)^*$; in matrix notation, $O_{ij} = O^*_{ji}$.

3. The Poisson bracket between two observables $\{O_1, O_2\}$ (Sect. 2.5.2:(P3)) is replaced by the *commutator* of the corresponding operators by the rule

$$\{O_1, O_2\} \rightarrow \frac{1}{\imath\hbar}\left[\hat{O}_1, \hat{O}_2\right] = \frac{1}{\imath\hbar}\left(\hat{O}_1\hat{O}_2 - \hat{O}_2\hat{O}_1\right), \tag{4.6}$$

with $\imath = \sqrt{-1}$ and $\hbar = h/2\pi$ is the reduced Planck constant.

4. For a system in state $|\psi\rangle \in \mathcal{H}$ the expectation value of an observable O at time t is given by

$$\langle O_t\rangle = \frac{\langle\psi|\hat{O}(t)|\psi\rangle}{\langle\psi|\psi\rangle} \tag{4.7}$$

5. For any physical state $|\psi\rangle \in \mathcal{H}$, there exist an operator for which $|\psi\rangle$ is one of its eigenstates,

$$\hat{Q}|\psi\rangle = q|\psi\rangle. \tag{4.8}$$

From the classical Poisson equation (Sect. 2.5.2(P3)) between the observable O and the time independent Hamiltonian function H

$$\frac{dO(t)}{dt} = \{O, H\}, \tag{4.9}$$

we get the quantum mechanical equation of motion in the *Heisenberg picture*

$$\frac{d\hat{O}(t)}{dt} = \frac{1}{\imath\hbar}\left[\hat{O}(t), \hat{H}\right], \tag{4.10}$$

which can be solved for a *time independent* Hamiltonian to yield

$$\hat{O}(t) = \exp(\imath\hat{H}t/\hbar)\hat{O}(0)\exp(-\imath\hat{H}t/\hbar). \tag{4.11}$$

Proof We define $U(t) = \exp(-\imath\hat{H}t/\hbar)$ and $U^+ = \exp(\imath\hat{H}t/\hbar)$, so

$$\hat{O}(t) = U^+\hat{O}(0)U. \tag{4.12}$$

We calculate

$$\begin{aligned}
\frac{d\hat{O}(t)}{dt} &= \frac{d}{dt}\left(U^+\hat{O}(0)U\right) \\
&= \frac{dU^+}{dt}\left(\hat{O}(0)U\right) + \left(U^+\hat{O}(0)\right)\frac{dU}{dt} \\
&= \frac{\imath\hat{H}}{\hbar}\left(U^+\hat{O}(0)U\right) - \left(U^+\hat{O}(0)\frac{\imath\hat{H}}{\hbar}U\right)
\end{aligned}$$

$$= \frac{1}{\hbar}\left[\hat{H}\left(U^{+}\hat{O}(0)U\right) - U^{+}\hat{O}(0)(UU^{+})\hat{H}U\right]$$

$$= \frac{1}{\hbar}\left[\hat{H}\left(U^{+}\hat{O}(0)U\right) - \left(U^{+}\hat{O}(0)U\right)\left(U^{+}\hat{H}U\right)\right]$$

$$= \frac{1}{\hbar}\left[\hat{H}\hat{O}(t) - \hat{O}(t)\hat{H}\right]$$

$$= \frac{1}{\hbar}\left[\hat{H}, \hat{O}(t)\right]$$

$$= \frac{1}{\imath\hbar}\left[\hat{O}(t), \hat{H}\right]$$

$$\square$$

The *unitary operator*, $U(t) = \exp(-\imath\hat{H}t/\hbar)$, is called *propagator* and it is used to transform from the *Heisenberg picture* with time dependent operators to the *Schrödinger picture* with time independent operators. Indeed, from the expectation value of the observable O at time t we get

$$\langle\psi|\hat{O}(t)|\psi\rangle = \langle\psi|e^{\imath\hat{H}t/\hbar}\hat{O}(0)e^{-\imath\hat{H}t/\hbar}|\psi\rangle$$
$$= \left[\langle\psi|e^{\imath\hat{H}t/\hbar}\right]\hat{O}(0)\left[e^{-\imath\hat{H}t/\hbar}|\psi\rangle\right]. \qquad (4.13)$$

If we write $|\psi(t)\rangle = e^{-\imath\hat{H}t/\hbar}|\psi\rangle$ we can see that the expectation value of $\hat{O}$ at time t is given as the expectation value of a time independent operator at time dependent states

$$\langle\psi|\hat{O}(t)|\psi\rangle = \langle\psi(t)|\hat{O}(0)|\psi(t)\rangle. \qquad (4.14)$$

Differentiating in time the equation $|\psi(t)\rangle = e^{-\imath\hat{H}t/\hbar}|\psi\rangle$ we get the *time dependent Schrödinger equation*

$$\imath\hbar\frac{\partial|\psi(t)\rangle}{\partial t} = \hat{H}|\psi(t)\rangle. \qquad (4.15)$$

The Schrödinger equation in the dual space takes the form

$$-\imath\hbar\frac{\partial\langle\psi(t)|}{\partial t} = \langle\psi(t)|\hat{H}. \qquad (4.16)$$

We define the *transition amplitude* in the Schrödinger picture as the *probability amplitude* for a particle to be found in the position x_f at time t_f from an initial position x_i at time t_i

$$K(x_f, t_f; x_i, t_i) = \langle x_f, t_f|x_i, t_i\rangle = \langle x_f|\exp\left(-\frac{1}{\hbar}\hat{H}(t_f - t_i)\right)|x_i\rangle. \qquad (4.17)$$

In a coordinate representation of quantum states, $|x_i, t_i\rangle$ denote the state vectors in Heisenberg picture,

$$\hat{q}|x_i, t_i\rangle = x_i|x_i, t_i\rangle, \qquad (4.18)$$

$|x_i\rangle$ are the state vectors in Schrödinger picture

$$\hat{q}|x_{i(f)}\rangle = x_{i(f)}|x_{i(f)}\rangle, \tag{4.19}$$

and $K(x_f, t_f; x_i, t_i)$ is called the *evolution operator* or *propagator*.

Therefore, the *transition probability density* from the initial position x_i at time t_i to the position x_f at time t_f is given by

$$P(x_f, t_f; x_i, t_i) = |K(x_f, t_f; x_i, t_i)|^2. \tag{4.20}$$

Spectral decomposition of the transition amplitude for time independent Hamiltonians is done by the following equations. Given

$$\hat{H}|n\rangle = E_n|n\rangle, \tag{4.21}$$

$$e^{i\hat{H}t/\hbar}|n\rangle = e^{iE_n t/\hbar}|n\rangle, \tag{4.22}$$

and

$$\sum_n |n\rangle\langle n| = \hat{I}, \tag{4.23}$$

then

$$
\begin{aligned}
K(x_f, t_f; x_i, t_i) &= \langle x_f| \exp\left(-\frac{1}{\hbar}\hat{H}(t_f - t_i)\right) |x_i\rangle \\
&= \langle x_f| \sum_n |n\rangle\langle n| \exp\left(-\frac{1}{\hbar}\hat{H}t_f\right) \exp\left(\frac{1}{\hbar}\hat{H}t_i\right) \sum_{n'} |n'\rangle\langle n'|x_i\rangle \\
&= \sum_n \langle x_f|n\rangle e^{-iE_n t_f/\hbar}\left(\sum_{n'} \langle n|n'\rangle e^{iE_{n'}t_i/\hbar}(\langle x_i|n'\rangle)^*\right) \\
&= \sum_n \psi_n(x_f)\psi_n^*(x_i)e^{-iE_n(t_f - t_i)/\hbar}. \tag{4.24}
\end{aligned}
$$

4.1.1 Quantum Hamilton's Equations

$\hat{q}$ is the coordinate operator with eigenvalues and eigenvectors $\hat{q}|x\rangle = x|x\rangle$ and normalized such that $\langle x|y\rangle = \delta(x - y)$, where $\delta(x - y)$ is Dirac's delta function.[3] The definition of delta function is

$$\delta(t) = \begin{pmatrix} \infty, & t = 0 \\ 0, & t \neq 0 \end{pmatrix} \tag{4.25}$$

---

[3] x designates the coordinate vector in a general coordinate system.

$$\int_{-\infty}^{\infty} \delta(t)dt = 1, \quad \text{and} \quad \int_{-\infty}^{\infty} f(t)\delta(t)dt = f(0). \qquad (4.26)$$

At this point it worth mentioning, that both δ-function and Gaussian function are employed to define initial distributions in phase space and they are related by the equation

$$\delta(t) = \lim_{\alpha \to 0} \frac{1}{\alpha\sqrt{\pi}} e^{-\frac{t^2}{\alpha^2}}. \qquad (4.27)$$

If we use as basis set the coordinate basis $|x\rangle$ the representation of the state $|\psi\rangle$ is

$$|\psi\rangle = \int_{-\infty}^{\infty} \psi(x)|x\rangle dx. \qquad (4.28)$$

The coefficient $\psi(x)$ is called *wavefunction* and it is a complex function, $\psi(x) \in \mathbb{C}$. Taking wavefunctions normalized to one

$$\langle\psi|\psi\rangle = \langle\psi| \left(\int_{-\infty}^{\infty} |x\rangle\langle x|dx \right) |\psi\rangle = \int_{-\infty}^{\infty} |\psi(x)|^2 dx = 1, \qquad (4.29)$$

we interpret $|\psi(x)|^2$ as the probability to find the particle in the interval $[x, x+dx]$. With the orthonormal coordinate basis the *completeness relation* is written

$$\int_{-\infty}^{\infty} |x\rangle\langle x|dx = \hat{I}. \qquad (4.30)$$

In the following we adopt the coordinate representation of the state vectors using wavefunctions in the Schrödinger picture, $\psi(x,t)$. We expand the wavefunction $\psi(x,t)$ of a dynamical system in an arbitrary complete basis set, $\chi_i(x)$, $i = 1, \ldots, n$

$$\psi(x,t) = \sum_{i=1}^{n} c_i(t)\chi_i(x). \qquad (4.31)$$

In this expansion the basis functions χ_i are time independent, whereas the coefficients c_i depend on time. ψ are solutions of the Schrödinger equation

$$i\hbar\frac{\partial\psi(x,t)}{\partial t} = \hat{H}\psi(x,t), \qquad (4.32)$$

and their complex conjugate the equation

$$-\imath\hbar\frac{\partial\psi^*(x,t)}{\partial t} = \psi^*(x,t)\hat{H}. \tag{4.33}$$

We assume ψ to be normalized, $\langle\psi(x,t)|\psi(x,t)\rangle = 1$, at any time. Then, by substituting Eq. 4.31 to the average value of the Hamiltonian, $\langle H\rangle$, we obtain

$$\langle H\rangle = \langle\psi|\hat{H}|\psi\rangle = \int_{-\infty}^{\infty}\psi^*\hat{H}\psi dx$$

$$= \sum_i\sum_j c_i^* c_j \int_{-\infty}^{\infty}\chi_i^*\hat{H}\chi_j dx. \tag{4.34}$$

We differentiate with respect to c_i^*

$$\frac{\partial\langle H\rangle}{\partial c_i^*} = \sum_j c_j\langle\chi_i|\hat{H}|\chi_j\rangle$$

$$= \langle\chi_i|\hat{H}|\psi\rangle$$

$$= \imath\hbar\frac{dc_i}{dt}. \tag{4.35}$$

Similarly, we take

$$\frac{\partial\langle H\rangle}{\partial c_j} = \sum_i c_i^*\langle\chi_i|\hat{H}|\chi_j\rangle$$

$$= \langle\psi|\hat{H}|\chi_j\rangle$$

$$= -\imath\hbar\frac{dc_j^*}{dt}. \tag{4.36}$$

We define the complex variables (Q^i, P_i) by introducing the real functions $q(t)$ and $p(t)$ to be assigned to the classical quantities of particle coordinates and conjugate momenta

$$Q^i(t) = c_i(t) = \frac{1}{\sqrt{2}}\left[q^i(t) + \imath p_i(t)\right]$$

$$P_i(t) = \imath\hbar c_i^*(t) = \frac{\imath\hbar}{\sqrt{2}}\left[q^i(t) - \imath p_i(t)\right]. \tag{4.37}$$

Eqs. 4.35 and 4.36 are the quantum equivalent of Hamilton's equations of motion

$$\dot{Q}^i = \frac{\partial}{\partial P_i}(\langle \psi|\hat{H}|\psi\rangle)$$

$$\dot{P}_i = -\frac{\partial}{\partial Q^i}(\langle \psi|\hat{H}|\psi\rangle). \tag{4.38}$$

We can take the inverse of Eq. 4.37 and write the real functions $q(t)$ and $p(t)$ as

$$q^i(t) = \frac{1}{\sqrt{2}}[c_i(t) + c_i^*(t)]$$

$$p_i(t) = \frac{-1}{\sqrt{2}}[c_i(t) - c_i^*(t)]. \tag{4.39}$$

4.1.2 Complexification of Classical Hamilton's Equations

Hamilton's equations in classical mechanics can also be defined in a complex manifold by *complexification* of phase space, i.e., by introducing the symplectic transformation

$$z = \frac{1}{\sqrt{2}}(q - \imath p)$$

$$w = \frac{1}{\sqrt{2}}(-\imath q + p)$$

$$= \imath \bar{z}, \tag{4.40}$$

where $\imath\bar{z}$ means the complex conjugate of $\imath z$. The inverse transformation is

$$q = \frac{1}{\sqrt{2}}(z + \imath w)$$

$$p = \frac{1}{\sqrt{2}}(\imath z + w)$$

$$= \imath \bar{z}. \tag{4.41}$$

For a harmonic oscillator in scaled normal coordinates (Sect. A.10) introducing these complex coordinates results in

$$H_0 = \frac{1}{2}\omega(Q^2 + P^2)$$

$$= \frac{1}{2}\omega\left[\frac{1}{2}\left(z^2 - w^2 + 2\imath zw\right) + \frac{1}{2}\left(-z^2 + w^2 + 2\imath zw\right)\right]$$

$$= \imath\omega zw. \tag{4.42}$$

Since, the transformation to complex coordinates and conjugate momenta is symplectic, Hamilton's equations are also written as

$$\dot{z} = \frac{\partial H'(z,w)}{\partial w}$$
$$\dot{w} = -\frac{\partial H'(z,w)}{\partial z},$$

(4.43)

where $H'(z, w) = H[Q(z, w), P(z, w)]$ is the complex Hamiltonian.

4.2 Quantum and Classical Autocorrelation Functions and Spectra

A spectrum is calculated by first evaluating the time autocorrelation function defined as

$$C(t) = \langle \psi(x, 0)| \psi(x, t) \rangle,$$

(4.44)

and then computing its Fourier transform,

$$I(E) = \frac{1}{2\pi\hbar} \int_{-\infty}^{\infty} \exp(\imath E t/\hbar) C(t)\, dt.$$

(4.45)

To see how the eigenvalues are extracted from Eq. 4.44, we expand $|\psi(x, t)\rangle$ as a series of the eigenfunctions $|n\rangle$ of the Hamiltonian $\hat{H}$ (Eq. 4.21)

$$|\psi(x, t)\rangle = \exp(-\imath \hat{H} t/\hbar)|\psi(x, 0)\rangle = \sum_n \exp(-\imath E_n t/\hbar)|n\rangle \langle n|\psi(x, 0)\rangle. \quad (4.46)$$

By introducing the overlap integral, $c_n = \langle n|\psi(x, 0)\rangle$, the spectrum becomes

$$I(E) = \frac{1}{2\pi\hbar} \sum_n |c_n|^2 \int_{-\infty}^{\infty} \exp[-\imath(E_n - E)t/\hbar]dt$$

$$= \frac{1}{2\pi\hbar} \sum_n |c_n|^2 \lim_{T\to\infty} \int_{-T}^{T} \exp[-\imath(E_n - E)t/\hbar]dt$$

$$= \frac{1}{2\pi} \sum_n |c_n|^2 \lim_{T\to\infty} \frac{2\sin[(E_n - E)T/\hbar]}{E_n - E}$$

$$= \sum_n |c_n|^2 \delta(E_n - E).$$

(4.47)

Thus, for infinite integration time (absolute resolution) we have a sum of delta functions located at the eigenvalues E_n. Finite integration in time (low resolution) will yield a sum of broaden peaks which may cover several E_n.

We could also extract the eigenfunctions $|n\rangle$ by computing the Fourier transform

$$|n\rangle = \frac{1}{c_n} \int_{-\infty}^{\infty} \exp(\imath E_n t/\hbar)|\psi(x,t)\rangle dt. \qquad (4.48)$$

It is clear, that in a time dependent calculation, we locate those eigenfunctions which overlap significantly with the initial wavefunction $|\psi(x,0)\rangle$. Since, we expect localization of the eigenfunctions at certain regions of configuration manifold, we simulate the spectrum by taking the appropriate initial wavepacket centered at the place of interest. It can be either an experimental spectrum or a theoretical one, which will reveal those states that are localized at particular regions of phase space.

Although quantum mechanical calculations are always what we should ask, however, they are not feasible at present for many degrees of freedom systems. It is useful then, to find the classical analogue of the quantum spectrum. In this case, the correspondence between spectrum and phase space structure is more straightforward.

The following formulation has been used in the past [1, 14, 15, 20]. By taking the square of the absolute value of the correlation function $C(t)$ (*the survival probability function*)

$$|C(t)|^2 = |\langle\psi(x,0)|\psi(x,t)\rangle|^2 = \langle\psi(x,0)|\psi(x,t)\rangle\langle\psi(x,t)|\psi(x,0)\rangle, \qquad (4.49)$$

and introducing the identity relation,

$$\hat{I} = \sum_n |n\rangle\langle n| \qquad (4.50)$$

we get,

$$\begin{aligned} |C(t)|^2 &= \langle\psi(x,0)|\psi(x,t)\rangle\langle\psi(x,t)|\hat{I}|\psi(x,0)\rangle \\ &= \sum_n [\langle n|\psi(x,0)\rangle\langle\psi(x,0)|\psi(x,t)\rangle\langle\psi(x,t)|n\rangle] \\ &= tr[\hat{\rho}(0)\hat{\rho}(t)]. \end{aligned} \qquad (4.51)$$

$\hat{\rho}(0)$ is the density operator,

$$\hat{\rho}(0) = |\psi(x,0)\rangle\langle\psi(x,0)|$$

and

$$\hat{\rho}(t) = |\psi(x,t)\rangle\langle\psi(x,t)| = e^{-\imath\hat{H}t/\hbar}|\psi(x,0)\rangle\langle\psi(x,0)|e^{\imath\hat{H}t/\hbar}, \qquad (4.52)$$

is the Heisenberg representation of $\hat{\rho}(0)$.

We can pass to the classical analogue by replacing the trace in Eq. 4.51 with an integral over the phase space, and by replacing the density operators with classical distribution functions,

$$\Omega(t) = \int \rho[q(0), p(0)]\rho[q(t), p(t)]dqdp. \tag{4.53}$$

The classical initial distribution $\rho(0)$ is usually a Wigner or a Husimi transform of the initial quantum wavefunction, $|\psi(q, 0)\rangle$, [24, 32]. The spectrum is then defined as the Fourier transform,

$$I_c(\omega) = \int e^{\imath \omega t} \Omega(t)dt. \tag{4.54}$$

If we count a conserved in time phase space distribution function , $g = \rho[q(t), p(t), t]$, as a dynamical variable, then, its time evolution is determined by solving Liouville's Equation (Sect. 2.5.2(P3))

$$\frac{d\rho}{dt} = \frac{\partial \rho}{\partial t} + \{\rho, H\} = 0, \tag{4.55}$$

where $\{ , \}$ is the Poisson bracket, H is the classical mechanical Hamiltonian and ρ the phase space density. Liouville's equation is stated as

$$\frac{\partial \rho}{\partial t} = -\{\rho, H\}. \tag{4.56}$$

4.3 Path Integral Quantum Mechanics

The probability to find a particle initially located at the position x_i the time t_i at the point x_f and later time t_f is given by the transition amplitude $\langle x_f, t_f | x_i, t_i \rangle$ in the Heisenberg picture. To convert it to the Schrödinger picture we use the Eq. 4.11

$$\hat{q}(t) = e^{\imath \hat{H}t/\hbar}\hat{q}(0)e^{-\imath \hat{H}t/\hbar} \equiv e^{\imath \hat{H}t/\hbar}\hat{q}e^{-\imath \hat{H}t/\hbar}, \tag{4.57}$$

and the relation between the position eigenvectors in the two pictures is

$$|x, t\rangle = e^{\imath \hat{H}t/\hbar}|x\rangle. \tag{4.58}$$

From the above equations we conclude that the transition amplitude in the Schrödinger picture is generated by

$$K(x_f, t_f; x_i, t_i) = \langle x_f, t_f | x_i, t_i \rangle = \langle x_f | e^{-\imath \hat{H}(t_f - t_i)/\hbar} | x_i \rangle. \tag{4.59}$$

Let us consider a Cartesian Hamiltonian

$$H = \frac{p^2}{2m} + V(x). \tag{4.60}$$

If the time interval $[t_i, t_f]$ is split in n small subintervals, such as $t_0 = t_i$ and $t_k = t_0 + \varepsilon k$ $(0 \leq k \leq n)$ and $t_n = t_f$, then, the transition amplitude takes the form

$$K(x_f, t_f; x_i, t_i) = \langle x_f, t_f | \int dx_{n-1} |x_{n-1}, t_{n-1}\rangle\langle |x_{n-1}, t_{n-1}|$$

$$\times \int dx_{n-2} |x_{n-2}, t_{n-2}\rangle\langle |x_{n-2}, t_{n-2}| \cdots$$

$$\times \int dx_1 |x_1, t_1\rangle\langle |x_1, t_1 | x_i, t_i\rangle, \tag{4.61}$$

where we have taken as $\varepsilon = (t_f - t_i)/n$ and applied the completeness relation, Eq. 4.30. As $n \to \infty$ we approximate

$$\langle x_k, t_k | x_{k-1}, t_{k-1}\rangle \approx \sqrt{\frac{m}{2\pi i\hbar\varepsilon}} e^{\frac{i}{\hbar}\Delta S_k}, \tag{4.62}$$

where

$$\Delta S_k = \varepsilon \left[\frac{m}{2} \left(\frac{x_k - x_{k-1}}{\varepsilon} \right)^2 - V\left(\frac{x_{k-1} + x_k}{2} \right) \right]. \tag{4.63}$$

Hence, we find

$$K(x_f, t_f; x_i, t_i) = \lim_{n\to\infty} \left(\frac{m}{2\pi i\hbar\varepsilon} \right)^{n/2} \int \exp\left(\frac{i}{\hbar} \sum_{k=1}^{n} \Delta S_k \right) \Pi_{j=1}^{n-1} dx_j. \tag{4.64}$$

We recognize that for a path from (x_i, t_i) to (x_f, t_f) and for $n \to \infty$ the quantity $\sum_{k=1}^{n} \Delta S_k$ is the action integral along this path

$$S[x(t)] = \int_{t_i}^{t_f} \left[\frac{m}{2} v^2 - V(x) \right] dt, \tag{4.65}$$

where v is the velocity. Since, x_k are considered as variables we interpret the integral 4.64 as a functional integral over all possible paths which connect the points (x_i, t_i) and (x_f, t_f) and it is symbolically written as

$$K(x_f, t_f; x_i, t_i) = \langle x_f, t_f | x_i, t_i \rangle = \int \exp\left[\frac{1}{\hbar} \int_{t_i}^{t_f} L\left(x, \frac{dx}{dt}\right) dt \right] \mathscr{D}x, \quad (4.66)$$

with L to symbolize the Lagrangian function. Equation 4.66 is the *path integral* formulation of the transition amplitude [9]. Notice, that the normalization factor in Eq. 4.64 has been incorporated in the symbol $\mathscr{D}x$.

For paths which start and end at the same configuration point, $x_f = x_i = x$, the path integral for trajectories over imaginary time can be identified with the quantum statistical *canonical partition function* of the system and with Hamiltonian that of Eq. 4.60. Indeed, if we take a *Wick rotation*, $t = -\imath\tau$, then

$$\frac{dx}{dt} = \imath\frac{dx}{d\tau}, \quad (4.67)$$

$$\exp(-\imath\hat{H}t/\hbar) = \exp(-\hat{H}\tau/\hbar), \quad (4.68)$$

$$\frac{1}{\hbar} \int_{t_i}^{t_f} \left[\frac{m}{2}\left(\frac{dx}{dt}\right)^2 - V(x) \right] dt = -\frac{1}{\hbar} \int_{\tau_i}^{\tau_f} \left[\frac{m}{2}\left(\frac{dx}{d\tau}\right)^2 + V(x) \right] d\tau.$$

The canonical partition function is written in coordinate representation as the *trace* of the operator $e^{-\imath\hat{H}(t_f - t_i)/\hbar} = e^{-\beta\hat{H}}$ as

$$Z(\beta) = \int \langle x | e^{-\beta\hat{H}} | x \rangle dx, \quad (4.69)$$

where we have defined the imaginary time $\beta = \imath(t_f - t_i)/\hbar$. For a Hamiltonian written in Cartesian coordinates the partition function becomes

$$Z(\beta) = \int_{closed\_paths} \exp\left\{ -\int_0^\beta \left(\frac{m}{2}\left(\frac{dx}{d\tau}\right)^2 + V(x) \right) \right\} \tilde{\mathscr{D}}x, \quad (4.70)$$

where the integral is over all closed paths in $[0, \beta]$.

Evaluating the trace of the propagator in a basis of energy eigenstates $|\psi_n\rangle$, $\hat{H}|\psi_n\rangle = E_n|\psi_n\rangle$, the partition function is written

$$Z(\beta) = \sum_n \exp(-\beta E_n), \quad (4.71)$$

where E_n are the eigenenergies of the Hamiltonian operator. Hence, we may conclude that if we are able to calculate the path integral over closed orbits and for time independent Hamiltonians, then, we can extract the energy spectrum.

4.4 Semiclassical Approximation

4.4.1 Gutzwiller's Periodic Orbit Semiclassical Quantization

Semiclassical theory can be formulated by applying the *stationary phase approximation (SPA)* (also known as the *saddle point approximation*) to the path integral expression of transition amplitude. We have seen that the propagator is written

$$K(x_f, t_f; x_i, t_i) = \langle x_f, t_f | x_i, t_i \rangle$$

$$= \int \exp \left[\frac{1}{\hbar} \int_{t_i}^{t_f} L\left(x, \frac{dx}{dt}\right) dt \right] \mathscr{D}x$$

$$= \int \exp \left[\frac{1}{\hbar} S[x] \right] \mathscr{D}x, \tag{4.72}$$

where $L(x, \dot{x}) = T - V$ is the Lagrangian function.

The basic tenet of the stationary phase approximation of such integrals is that for small $\hbar$ ($\hbar \to 0$), the integrand oscillates so rapidly that the integral over any small x-interval will yield zero unless one is close to a *stationary point* x_c of $S[x]$, $\delta S[x_c] = 0$, for which to first order around x_c there are no oscillations. This suggests that for $\hbar \to 0$ the integral is dominated by the contribution from the neighbourhood of some stationary points x_c of $S[x]$, and that therefore in this limit the dominant contribution to the integral can be obtained by a Taylor expansion of $S[x]$ around $x = x_c$.

Thus, SPA for the evaluation of the complex integral in Eq. 4.72 involves,

1. first the calculation of the stationary points of the integrand

$$\delta S[x_f, x_i] = \delta \int_{t_i}^{t_f} L\left(x(t), \frac{dx(t)}{dt}\right) dt = 0, \tag{4.73}$$

2. and second, expansion of the integrand to a Taylor series around the stationary points, usually up to the second order

$$S(x_c + \delta x) \approx S(x_c) + \delta S(x) + \frac{1}{2} \delta^2 S(x) + \cdots . \tag{4.74}$$

For the integral curves of the Hamiltonian flow with fixed initial and final coordinates $[x_f, x_i]$, it is valid $\delta S(x) = 0$. Under these approximations the propagator takes the form [12, 18]

$$\langle x_f | \exp(-\imath \hat{H} t / \hbar) | x_i \rangle = \sum_{roots} \left[(2\pi \imath \hbar)^n \left| \frac{\partial x_f}{\partial p_i} \right| \right]^{-1/2}$$

$$\times \exp\left[\imath \left(S(x_f, x_i, t) / \hbar - \mu \pi / 2 \right) \right]. \quad (4.75)$$

This is the *semiclassical approximation* to the transition amplitude (Van Vleck [18]). The sum is over all root of Eq. 4.73. μ is the *Maslov index* and it depends on the number of turning points in oscillatory motion. $\hat{H}$ is the Hamiltonian operator for a system with n degrees of freedom. $S(x_f, x_i, t)$ is the action along the trajectories from the initial configuration point x_i to the point x_f arriving at the time interval t, and p_i the conjugate momentum of x_i. To find the roots of Eq. 4.73, one must solve a nonlinear boundary problem. As it is expected, this involves linearization of the equations of motion and computation of Jacobi fields.

We have shown, that the path integral formulation of the propagator can take the expression of the canonical partition function if we adopt closed paths, Eq. 4.70. We can further show that the semiclassical approximation of the transition amplitude involves periodic orbits in phase space. Invoking the stationary phase approximation for trajectories returning to the initial configuration ($x_f = x_i = x$) in the time period T

$$\frac{\partial S(x, x, T)}{\partial x} = 0 \quad (4.76)$$

and recalling Eqs. 2.43 and 2.44, we find that the major contributions to the integral in Eq. 4.75 come from the periodic orbits [18]

$$\left[\frac{\partial S(x_f, x_i, T)}{\partial x_i} + \frac{\partial S(x_f, x_i, T)}{\partial x_f} \right]_{x_i = x_f = x} = -p_i + p_f = 0. \quad (4.77)$$

This result signifies the importance of periodic orbits in computing molecular spectra. Furthermore, we have already seen that the scarring theory of Heller [19] demonstrated that the eigenfunctions may stay localized around unstable periodic orbits. It turns out, that the *scarring* of the wavefunctions by stable or the least unstable periodic orbits is a general phenomenon in polyatomic molecules.

4.4.2 The Semiclassical Trace Formula

The fluctuating part, $N(E)$, of the density of states, $D(E)$, for a bound system with n degrees of freedom and at energy E, is predicted by Gutzwiller's trace formula [18],

$$D(E) = \sum_{n} \delta(E - E_n) = D_{av}(E) + N(E), \quad (4.78)$$

where

$$N(E) = Re \sum_{po} \frac{T_{po}}{\pi\hbar} \sum_{j=1}^{\infty} \frac{\exp[\imath j (S_{po}/\hbar - \mu_{po}\pi/2)]}{[det(M_{po}^j - 1)]^{1/2}}. \qquad (4.79)$$

$D_{av}(E)$ designates the average density of states and $\imath = \sqrt{-1}$. The summation in Eq. 4.79 is over all periodic orbits with period T_{po}, action S_{po}, and for an infinite number of loops j. μ_{po} is the Maslov index and M_{po}^j is the linear stability matrix (monodromy matrix), which describes the results of the transverse displacements off the periodic orbit after j loops. The determinant in Eq. 4.79 is evaluated from the eigenvalues of the monodromy matrix, Eq. 3.39,

$$det(M_{po}^j - 1) = \prod_{i=1}^{2(n-1)} (\lambda_{po}^j(i) - 1), \qquad (4.80)$$

where $\lambda_{po}^j(i)$ is the jth power of the ith eigenvalue (a complex or real number) of the monodromy matrix which corresponds to a perpendicular direction of the periodic orbit . All the above quantities are computed as functions of the total energy during the periodic orbit analysis.

Generalizations of the trace formula have been given [2, 3], and Miller [26] has proved a relation for semiclassically quantizing specific stable periodic orbits. According to Eq. 4.79, the calculation of the density of states requires a summation over all periodic orbits and for an infinite number of loops of each periodic orbit. This, of course, is impossible for real systems, and therefore there are convergence problems with this equation. Nevertheless, applications of the trace formula, such as the hydrogen atom in strong magnetic fields [11], have shown that satisfactory results may be obtained by employing a finite number of periodic orbits. Although, accurate eigenvalues is difficult to achieve, low resolution characteristics of the spectra can be identified, and thus, a connection of particular periodic orbits with spectral peaks to be made [29].

4.4.3 Einstein-Brillouin-Keller Quantization Rule for Integrable Systems

For integrable systems of n degrees of freedom the Einstein-Brillouin-Keller semi-classical quantization is applied. Integrability implies the existence of n-action variables, which are quantized according to the equation [8]

$$\frac{1}{2\pi} \oint_{C_i} p_i dq^i = \left(m_i + \frac{1}{4}\mu_i\right)\hbar, \quad i = 1, \ldots, n, \ m_i = 0, 1, 2, \ldots, \qquad (4.81)$$

where p_i are the conjugate momenta to the coordinates q^i, C_i the n topologically independent closed integration curves, and μ_i the Maslov index (grossly speaking the number of turning points in the ith coordinate).

4.4.4 Initial Value Representation Semiclassical Quantization

Miller [27] has proposed a semiclassical method appropriate for polyatomic molecules studied by classical dynamics. The latter mainly involves Hamiltonians formulated in Cartesian coordinates. The name of this method is *Initial Value Representation* (IVR) to emphasize that the semiclassical formula for the transition amplitude can be written by an equation which requires the solution of an initial value problem than of a nonlinear boundary problem as is more often met in semiclassical theories.

Writing the probability amplitude for a transition from the state ψ_{n_1} at time zero to state ψ_{n_2} at time t requires averaging over the initial (x_1) and final (x_2) coordinates

$$
K_{n_2,n_1}(t) = \langle \psi_{n_2} | \exp(-\imath \hat{H} t/\hbar | \psi_{n_1} \rangle
$$
$$
= \int dx_1 \int dx_2 \psi^*_{n_2}(x_2) \psi_{n_1}(x_1) \langle x_2 | \exp(-\imath \hat{H} t/\hbar) | x_1 \rangle. \quad (4.82)
$$

By employing the semiclassical approximation for the matrix elements $\langle x_2 | \exp(-\imath \hat{H} t/\hbar) | x_1 \rangle$, i.e., the Van Vleck approximation, Eq. 4.75, the transition amplitude becomes

$$
K_{n_2,n_1}(t) = \sum_{roots} \int dx_1 \int dx_2
$$
$$
\times \psi^*_{n_2}(x_2) \psi_{n_1}(x_1) \left[(2\pi\imath\hbar)^n \left| \frac{\partial x_2}{\partial p_1} \right| \right]^{-1/2}
$$
$$
\times \exp \left[\imath \left(S(x_2, x_1, t)/\hbar - \mu\pi/2 \right) \right]. \quad (4.83)
$$

The initial-value-representation "trick" is based on realizing that the Van Vleck determinant, $\left| \frac{\partial x_2}{\partial p_1} \right|$, is nothing else but the Jacobian of the transformation from the position x_2 to the initial momentum p_1. Thus, by replacing the summation over the roots and the integration over x_2 by

$$
\sum_{roots} \int dx_2 \Longrightarrow \int dp_1 \left| \frac{\partial x_2}{\partial p_1} \right|, \quad (4.84)
$$

we obtain the IVR formula of the transition amplitude

$$K_{n_2,n_1}(t) = \int dx_1 \int dp_1$$

$$\times \psi_{n_2}^*(x_t)\psi_{n_1}(x_1) \left[(2\pi i\hbar)^n \left| \frac{\partial x_t}{\partial p_1} \right| \right]^{+1/2}$$

$$\times \exp\left[i\left(S(x_t, x_1, t)/\hbar - \mu\pi/2 \right) \right]. \tag{4.85}$$

We see, that the nonlinear boundary value problem has been replaced by an initial value one, which is much simpler since for each initial position and momentum, there is only one classical trajectory. Moreover, the singularities of the Van Vleck determinant have been replaced by zeroes of its inverse. The initial value representation can describe classically forbidden processes such as barrier or dynamical tunnelling, using real-valued trajectories.

Comparing Eqs. 4.79 and 4.85 we immediately see the similarities as both formulae are founded on the semiclassical approximation of the transition amplitude. The trace formula for the density of states includes summation over an infinity number of specific orbits (periodic), whereas the application of IVR formula requires sampling over random trajectories initialized from a predetermined initial distribution. Nevertheless, the convergence difficulties encountered in both formulae arise from the oscillatory imaginary term. Miller and coworkers have proposed several ingenious 'tricks' to meliorate the semiclassical approximation and the successful applications of IVR approach make this semiclassical method quite promising for polyatomic molecules [27, 28]. However, it is fair to admit that for small molecules (triatomic/tetratomic) is simpler to solve the Schrödinger equation than trying to converge semiclassical approximations.

4.5 Solving Schrödinger Equation in Cartesian Coordinates via Angular Momentum Projection Operators

Accurate Quantum Molecular Dynamics (QMD), which require the numerical solution of the Schrödinger equation [23], apart from providing physical insight into the nature of the processes, are also imperative for studying quantum effects such as tunnelling and interferences in chemical dynamics and spectroscopy. Moreover, they produce quantitative results that can be used as benchmark for other approximate approaches, such as classical and semiclassical methods suitable for large systems [27].

Variational methods usually approximate the solutions of Schrödinger equation assuming considerable knowledge about the system; this can become especially awkward for large and complex molecules. There is also the problem of choosing the appropriate basis set of functions on which the process will be described, and the subsequent check of its completeness. On the other hand, a time dependent representation simplifies the interpretation of the dynamics as well as the identification of the degrees of freedom involved in the computation, as it was explained in [21].

Hence, much effort is being devoted to developing new techniques for solving the Schrödinger equation (time dependent and time independent) for realistic molecular systems. Nevertheless, their numerical implementation requires a considerable computational load.

Solving numerically the Schrödinger equation, we must take into account that the sparsity of the Hamiltonian matrix is increased by using the so called *local* methods, where local means that the action of the Hamiltonian operator on the wavefunction is carried out using information from the surrounding coordinate points. To this end, Variable Order Finite Difference method (VOFD) [13, 16, 17] for approximating the second order derivatives in the kinetic energy operator is a versatile and widely applied method. In Chap. 5 we describe the VOFD method. However, before developing the new algorithms for QMD, one must first decide on the most convenient and general coordinate system.

The standard way of describing a polyatomic molecule with N atoms in the absence of external fields is to introduce the center of mass, three Euler angles and $3N - 6$ internal coordinates (frame transformation) [22]. The problem with this methodology is twofold. First, there is the analytical derivation of the Hamiltonian operator in curvilinear coordinates and Euler angles or angular momentum operators. Not only this is a cumbersome task for more than three atom molecules [5], but this procedure has to be repeated for every choice of curvilinear coordinates and angles necessary for studying different molecules and phenomena. Second, the appearance of cross derivatives in the kinetic term of the Hamiltonian implies a computational effort that scales as N^2.

An alternative theoretical approach can be devised for avoiding these limitations and easily extending the applications to polyatomic molecules by not separating rotations at the operator level but to use projection techniques to conserve the total angular momentum. The total translation of the molecule can easily be eliminated by choosing $N - 1$ Jacobi vectors [25]. This yields a $3N - 3$ Cartesian coordinate system which results in a diagonal and generic formula for the kinetic energy operator compared to the standard curvilinear coordinate procedures, as was shown in [30]. This part will be explained in detail in the next section where the computational methods are described.

4.5.1 Computational Methods

If the vector x denotes the $3N-3$ mass-weighted Cartesian coordinates of a molecule with N-atoms, the time dependent Schrödinger equation (TDSE) is written as

$$\hat{H}\psi(x,t) = \left[-\frac{\hbar^2}{2} \sum_{i=1}^{n} \frac{\partial^2}{\partial x_i^2} + V(x) \right] \psi(x,t) = \imath\hbar \frac{\partial \psi(x,t)}{\partial t}, \qquad (4.86)$$

where $\imath = \sqrt{-1}$, $\hat{H}$ the Hamiltonian operator and $\psi(x, t)$ is the wavefunction at time t. On the left side of the equation we find the kinetic and potential energy operators. The potential function, V, is assumed to be time independent. Equation 4.86 has then the formal solution

$$\psi(x, t) = U(t, \hat{H})\psi_0(x) = \exp(-\imath \hat{H}t/\hbar)\psi_0(x), \tag{4.87}$$

where $U(t, \hat{H}) = \exp(-\imath \hat{H}t/\hbar)$ is the time propagator and $\psi_0(x)$ is the initial wavefunction (wavepacket) at $t = 0$.

The proposed algorithm for solving Eq. 4.86 comprises four steps. (I) We discretize the wavefunction on a grid of points in the Cartesian coordinate space, then, we approximate the wavefunction over these grid points with Lagrange interpolation polynomials [10]. The latter allows us to compute the second derivatives of ψ. (II) The projection of the initial wavefunction, ψ_0, at a specific irreducible representation subspace of the total angular momentum, J, of dimension $2J + 1$ [34] is taken. (III) We evaluate the action of the Hamiltonian operator on the projected wavefunction ψ_J. (IV) The propagation of the wavefunction in time by means of an extrapolation procedure is executed.

4.5.2 The Angular Momentum Projection Operators

For a specified total angular momentum, J, we assume that a projection operator, $\hat{\pi}_J$, exists that projects out the wavefunction,

$$\psi_J(x) = \hat{\pi}_J \psi(x), \tag{4.88}$$

onto the subspace spanned by all eigenfunctions of the J_z component of angular momentum, each assigned by the quantum number K $(-J \leq K \leq J)$. Thus, one has to find projection operators, $\hat{\pi}_J$, which are idempotent, Hermitian, and commute with $\hat{H}$ due to the rotational invariance of the Hamiltonian, i.e.,

$$(\hat{\pi}_J)^2 = \hat{\pi}_J, \quad (\hat{\pi}_J)^\dagger = \hat{\pi}_J, \quad [\hat{H}, \hat{\pi}_J] = 0. \tag{4.89}$$

Equivalently to the projection of ψ to ψ_J, we may also consider the Hamiltonian $\hat{H}_J = \hat{\pi}_J \hat{H} \hat{\pi}_J$ acting on ψ. This is the operator that is derived in frame transformation theory for every choice of internal coordinates, Euler angles and total angular momentum. We now show below that rotation group theory can avoid its explicit derivation [6].

Considering that the rotations about the center of mass are described by the three Euler angles, (ϕ, θ, γ), then, the operators $R(\Omega) = e^{-\imath \phi J_z} e^{-\imath \theta J_y} e^{-\imath \gamma J_z}$, constitute the elements of the rotation group of the system, while the associated irreducible representations of the group are provided by the Wigner functions [33]

$$D_{MK}^J(\Omega) = e^{-\imath M\phi} d_{MK}^J(\theta) e^{-\imath K\gamma}. \tag{4.90}$$

From the group shift operators

$$\hat{\pi}_{JMK} = \frac{2J+1}{8\pi^2} \int d\Omega\, D_{MK}^{*J}(\Omega) R(\Omega), \tag{4.91}$$

and their properties

$$\sum_J \hat{\pi}_J = 1, \quad \hat{\pi}_J = \sum_K \hat{\pi}_{JKK}, \tag{4.92}$$

we can construct the projection operator onto the space of the total angular momentum J as the K sum of the $\hat{\pi}_{JKK}$ operators [6, 7].

We can deduce several properties of the molecule from the solutions of time dependent Schrödinger equation by introducing autocorrelation functions. Many molecular properties such as spectra and cross sections can be obtained by Fourier transforming the time correlation function for a specific angular momentum J

$$C_J(t) = \langle \psi(0)|\hat{\pi}_J \exp(-\imath \hat{H}t/\hbar)\hat{\pi}_J|\psi(0)\rangle = \langle \psi(0)|\exp(-\imath \hat{H}t/\hbar)|\hat{\pi}_J \psi(0)\rangle. \tag{4.93}$$

This expression is the key to the method: introducing the angular momentum is equivalent to propagating a projection of the total wavefunction onto the J subspace, which is obtained from

$$\psi_J(x) = \hat{\pi}_J \psi_0(x) = \frac{2J+1}{8\pi^2} \int d\Omega \sum_K D_{KK}^{*J}(\Omega)\, R(\Omega)\psi_0(x). \tag{4.94}$$

The projection of the initial state is carried out only once, at the beginning of the calculations, and it is normally, much faster than the time propagation step. Notice, that the cumbersome projected Hamiltonian $\hat{H}_J$ does not appear in the autocorrelation function of Eq. 4.93. In the next chapter we describe all the steps to solve the TDSE by discretizing a Cartesian coordinate space.

References

1. Baranger M (1958) Simplified quantum-mechanical theory of pressure broadening. Phys Rev 111:481–493
2. Berry MV, Tabor M (1977) Level clustering in the regular spectrum. Proc Royal Soc Lond A 356:375
3. Bogomolny WB (1992) On dynamical zeta function. Chaos 2:5
4. Bohigas O, Giannoni MJ, Schmidt C (1986) In: Quantum chaos and statistical nuclear physics, Lecture Notes in Physics, vol 263. Springer
5. Bramley MJ, Handy NC (1981) Efficient calculation of rovibrational eigenstates of sequentially bonded four-atom molecules. J Chem Phys 98(2):1378–1397

6. Broeckhove J, Lathouwers L (1993) Quantum dynamics and angular momentum projection. In: Cerjan C (ed) Numerical grid methods and their applications to Schrödinger equation. Kluwer Academic Publishers, Dordrecht, pp 49–56

7. Corbett JO (1971) A note on angular momentum projection operators. Nucl Phys A 169: 426–428

8. Curtis LJ, Ellis DG (2004) Use of the Einstein-Brillouin-Keller action quantization. Am J Phys 72:1521–1523

9. Feynman RP, Hibbs AR (1965) Quantum mechanics and path integrals. McGraw-Hill, New York

10. Fornberg B (1998) A practical guide to pseudospectral methods. Cambridge University Press, Cambridge

11. Friedrich H, Wintgen D (1989) The Hydrogen atom in a uniform magnetic field - an example of chaos. Phys Rep 183:37

12. Gaspard P, Burghardt I (1997) Emergence of classical periodic orbits and chaos in intramolecular and dissociation dynamics. Adv Chem Phys 101:491–621

13. Goldfield EM, Gray SK (2007) Quantum dynamics of chemical reactions. Adv Chem Phys 136:1–37

14. Gomez Llorente JM, Taylor HS (1989) Spectra in the chaotic region: a classical analysis for the sodium trimer. J Chem Phys 91:953–962

15. Gomez Llorente JM, Pollak E (1992) Classical dynamics methods for high energy vibrational spectroscopy. Annu Rev Phys Chem 43:91–126

16. Guantes R, Farantos SC (1999) High order finite difference algorithms for solving the Schrödinger equation in molecular dynamics. J Chem Phys 111:10,827–10,835

17. Guantes R, Farantos SC (2000) High order finite difference algorithms for solving the Schrödinger equation in molecular dynamics. II. periodic variables. J Chem Phys 113:10429–10437

18. Gutzwiller MC (1990) Chaos in Classical and Quantum Mechanics, vol 1. Springer, New York

19. Heller EJ (1984) Bound-state eigenfunctions of classically chaotic Hamiltonian systems: scars of periodic orbits. Phys Rev Lett 53(16):1515–1518

20. Heller EJ, Davis MJ (1980) Molecular overtone bandwidths from classical trajectories. J Phys Chem 84:1999–2000

21. Kosloff R (1996) Quantum molecular dynamics on grids. In: Wyatt RE, Zhang J (eds) Dynamics of molecules and chemical reactions. Marcel Dekker Inc, New York, pp 185–230

22. Landau LD, Lifshitz EM (1976a) Mechanics, 3rd edn. Pergamon Press, Oxford

23. Landau LD, Lifshitz EM (1976b) Quantum mechanics, 3rd edn. Pergamon Press, Oxford

24. McDonald SW (1988) Phase-space representations of wave equations with applications to the eikonal approximation for short-wavelength waves. Phys Rep 158:337

25. Meyer KR, Hall GR, Offin D (2009) Introduction to Hamiltonian dynamical systems and the N-Body problem, vol 90, 2nd edn., Applied mathematical sciences.Springer, New York

26. Miller WH (1975) Semiclassical quantization of nonseparable systems: a new look at periodic orbit theory. J Chem Phys 63:996

27. Miller WH (2001) The semiclassical initial value representation: a potentially practical way for adding quantum effects to classical molecular dynamics simulations. J Phys Chem A 105:2942–2955

28. Miller WH (2009) Electronically nonadiabatic dynamics via semiclassical initial value methods. J Phys Chem A 113:1405–1415

29. Prosmiti R, Farantos SC, Taylor HS (1994) A periodic orbit approach to spectroscopy and dynamics of $SO_2:\tilde{C} \, ^1B_2 \rightarrow \tilde{X} \, ^1A_1$. Mol Phys 82:1213–1232

30. Suarez J, Farantos SC, Stamatiadis S, Lathouwers L (2009) A method for solving the molecular Schrödinger equation in Cartesian coordinates via angular momentum projection operators. Comput Phys Commun 180:2025–2033

31. Waterland RL, Yuan JM, Martens CC, Gillilan RE, Reinhardt WP (1988) Classical-quantum correspondence in the presence of global chaos. Phys Rev Lett 61:2733

32. Weissman Y, Jortner J (1982) Quantum manifestations of classical stochasticity. II. dynamics of wave packets of bound states. J Chem Phys 77:1486
33. Wigner EP (1959) Group theory. Academic Press, London
34. Zare RN (1988) Angular momentum, understanding spatial aspects in chemistry and physics. Wiley-Interscience, New York

Chapter 5
Numerical Methods

The theories developed in the previous chapters, classical and quantum mechanical, are put in action by discretizing the corresponding differential equations. The variable order finite difference approximations to the unknown solutions and their derivatives are the preferred methods, not only because of their well understood convergence properties and the relatively easy way of their programming, but also, finite differences provide a uniform approach to the different type of equations, especially when we work in a Cartesian coordinate system. With respect to Schrödinger equation several grids are examined and comparisons with the more popular pseudospectral methods is made. For the location of periodic orbits the multiple shooting method is developed as it has been thoroughly tested. Finally, computer codes for studying classical nonlinear molecular dynamics and solving the Schrödinger equation are described.

5.1 Discretizing the Schrödinger Equation

5.1.1 Variable Order Finite Difference Methods

Expanding functions in a series of polynomials is a common practise to find solutions of differential equations as well as representing molecular potential as Taylor series around equilibrium points. Therefore, robust numerical methods to compute high order derivatives are required and finite difference is one of them. Finite Difference (FD) approximations of the derivatives of a function $F(x)$ can be extracted by interpolating $F(x)$ with Lagrange polynomials, $P_n(x)$. This allows one to calculate the derivatives analytically at arbitrarily chosen grid points and with a *variable order* of approximation.

© The Author(s) 2014
S.C. Farantos, *Nonlinear Hamiltonian Mechanics Applied to Molecular Dynamics*,
SpringerBriefs in Electrical and Magnetic Properties of Atoms, Molecules, and Clusters,
DOI 10.1007/978-3-319-09988-0_5

Let N to be the total number of interpolation points and n a subset of it employed for the interpolation of the function $F(x)$. The *Lagrange fundamental polynomials of order n* are defined by

$$L_k^n(x) = \prod_{j=0}^{n}{}' (x - x_j) / \prod_{j=1}^{n}{}' (x_k - x_j), \quad n = 0, 1, \dots, N, \tag{5.1}$$

where the prime means that the term $j = k$ is not included in the products. The values of $L_k^n(x_j)$ are zero for $j \neq k$ and one for $j = k$ by construction, i.e., $L_k^n(x_j)$ consist a *cardinal set* of basis functions

$$L_k^n(x_j) = \delta_{j,k}, \quad 0 \leq j, k \leq N, \tag{5.2}$$

and

$$\delta_{ij} = 1, \quad i = j, \\ = 0, \quad i \neq j. \tag{5.3}$$

The function can then be approximated as

$$F(x) \approx P_n(x) = \sum_{k=0}^{n} L_k^n(x) F(x_k). \tag{5.4}$$

P_n is a polynomial of order n. The coefficients in Eq. 5.4 are simply the values of the function F at the *collocation points* x_k.

The main requirements for the basis functions are these approximations, $F(x) \approx P_n(x)$, must converge rapidly to the true solution with the order of approximation n, and that, given the coefficients $F(x_k)$, the determination of $b_{j,k}^{(m)}$ such that

$$\frac{d^m}{dx^m} \left(\sum_{k=0}^{n} L_k^n(x) F(x_k) \right) \Bigg|_{x=x_j} = \sum_{k=0}^{n} b_{j,k}^{(m)} F(x_k), \quad n = 0, \dots, N, \tag{5.5}$$

should be efficient. Thus, we define

$$b_{j,k}^{(m)} = \frac{d^m L_k^n(x)}{dx^m} \Bigg|_{x=x_j}, \tag{5.6}$$

and by Taylor's theorem we can write $L_k^n(x)$ as

$$L_k^n(x) = \sum_{m=0}^{n} \frac{b_{j,k}^{(m)}}{m!} x^m. \tag{5.7}$$

Also, it must be possible to convert quickly between the coefficients $F(x_k)$ and the values of the sum $P_n(x_j)$ at the set of collocation points. Equation 5.5 for all the x_j can be expressed as a matrix-vector multiplication

$$\frac{d^m P_n(x)}{dx^m} = D^m \cdot v^T, \qquad (5.8)$$

where the *differentiation matrix* D^m contains the coefficients necessary for calculating the mth derivative at the collocation points and v^T is the column vector of dimension $N + 1$ containing the basis functions.

The functions $L_k^n(x)$ are obtained recursively by the equations

$$L_k^n(x) = \frac{x - x_n}{x_k - x_n} L_k^{n-1}(x), \qquad (5.9)$$

$$L_n^n(x) = \frac{\prod_{l=0}^{n-2}(x_{n-1} - x_l)}{\prod_{l=0}^{n-1}(x_n - x_l)}(x - x_{n-1})L_{n-1}^{n-1}(x), \qquad (5.10)$$

and initializing with

$$b_{0,0}^{(0)} = 1. \qquad (5.11)$$

With these iteration equations the matrix D^m is computed with the lth derivative, $l = 0, 1, \ldots, m$, and $k = l, l+1, \ldots, N$. A fast algorithm for calculating these coefficients for any order and arbitrarily spaced grid points has been produced by Fornberg [19]. We have successfully applied the above Lagrange interpolation scheme to molecular problems [22–24].

5.1.2 Pseudospectral Methods

Grid representations of the Schrödinger equation can be obtained by first defining global smooth and analytic basis functions, $\phi_j(x)$, to expand the wavefunction as

$$\psi(x) \approx \psi_N(x) = \sum_{j=1}^{N} a_j \phi_j(x). \qquad (5.12)$$

Different global basis functions define different Pseudospectral methods. From such so-called Finite Basis Representation (FBR) we can transform to a cardinal set of basis functions, $u_j(x)$, or the Discrete Variable Representation (DVR) as is usually called, by choosing N grid points, x_i, at which the wavefunction is calculated. The cardinal functions obey the δ-Kronecker property

$$u_j(x_i) = \delta_{ij}, \qquad (5.13)$$

so that the wavefunction is represented in the set of grid points as

$$\psi_N(x) = \sum_{j=1}^{N} \psi(x_j)u_j(x). \tag{5.14}$$

Notice, that the expansion coefficients are the exact wavefunction values, $\psi(x_j)$, at the chosen grid points.

The transformation from FBR to the cardinal basis set is unitary and the new basis is expressed in terms of the old one by:

$$u_j(x) = \sum_{i=1}^{N} <\phi_i|u_j> \phi_i(x), \quad j = 1, \ldots, N. \tag{5.15}$$

A common procedure now is to evaluate the matrix elements $<\phi_i|u_j>$ by Gaussian quadrature, such that the integral becomes exact for a polynomial type basis. A number of different approximation methods can be obtained by defining different basis functions and quadrature rules [7, 27, 37]. In general, using the δ-Kronecker property of $u_j(x)$ (Eq. 5.13) we can write

$$<\phi_i|u_j> = \sum_{k=1}^{N} w_k\phi_i^*(x_k)u_j(x_k) = w_j\phi_i^*(x_j), \tag{5.16}$$

where ϕ_i^* is the complex conjugate function of ϕ_i, and the grid points x_k and the corresponding weights w_k depend on the chosen quadrature rule.

5.1.2.1 Periodic Uniform Grids

Most of vibrational Hamiltonians or Hamiltonians describing molecule-surface encounters, atom-diatom or four-atom chemical reactions require the use of angular variables, and therefore, periodic boundary conditions. It is interesting to see if FD methods can be applied to angular variables with the same effectiveness demonstrated for radial variables and to investigate if the same limits can be approached here. In this context, it is worth studying FD approximations with different grid distributions. Doing this we can compare some local approaches to the solution of the Schrödinger equation with well established DVR methods used for angular variables, such as Legendre or Chebyshev orthogonal polynomial expansions which lead to non uniform grids compared to Fourier method which is based on uniform grids.

A widely used set of basis functions in the solution of Schrödinger equation is the Fourier set:

$$\phi_j(x) = \frac{1}{\sqrt{2\pi}} \exp(\imath 2\pi j x/L), \tag{5.17}$$

where L is the length which defines the periodicity of the function, $j = -M, \ldots, 0, \ldots, M$ and $\imath = \sqrt{-1}$. We can transform these functions to a set of cardinal functions by using a *uniform grid* in x with an *odd* number of points, $N = 2M + 1$, and employing Chebyshev quadrature. This results in the *real* trigonometric series [19],

$$u_j(x) = \frac{w_j}{2\pi}[1 + 2\sum_{k=1}^{M} \cos(2\pi k(x - x_j)/L)], \tag{5.18}$$

The weights are equal to $w_j = L/N$ for Chebyshev quadrature. Assuming for simplicity that x is an angular variable with period $L = 2\pi$, and using the trigonometric identity

$$\frac{1}{2} + \sum_{k=1}^{M} \cos(k\alpha) = \frac{\sin[(M + 1/2)\alpha]}{2\sin(\alpha/2)}, \tag{5.19}$$

Eq. 5.18 gives the *Fourier cardinal functions*

$$u_j(x) = \frac{\sin[N(x - x_j)/2]}{N\sin[(x - x_j)/2]}. \tag{5.20}$$

The δ-Kronecker property can be immediately checked, and we can derive analytically the derivatives of the wavefunction $\psi(x)$ when it is expanded in the cardinal functions $u_j(x)$ (Eq. 5.14):

$$\frac{d^m\psi(x)}{dx^m}\bigg|_{x=x_k} = \sum_{j=1}^{N} b_{k,j}^{(m)}\psi(x_j), \tag{5.21}$$

with $b_{k,j}^{(m)}$ the mth derivative of $u_j(x)$ evaluated at $x = x_k$.

In the next section we show how Fourier cardinal functions and Finite Differences are related. Fourier cardinal basis can be seen as a sum of Sinc basis functions repeated periodically with periodicity 2π, i.e.,

$$\frac{\sin[N(x - x_j)/2]}{N\sin[(x - x_j)/2]} = \sum_{k=-\infty}^{k=\infty} \text{Sinc}[(x - x_j + 2\pi k)/\Delta x], \tag{5.22}$$

$\Delta x = 2\pi/N$ in this case and we have used the fraction expansion $\pi/\sin(\pi x) = \sum_{k=-\infty}^{k=\infty}(-1)^k/(x + k)$ [21].

A cardinal basis function widely used in interpolation theory is the Sinc function [44], defined as $\text{Sinc}(x) \equiv \sin(\pi x)/\pi x$. The Sinc function can be naturally generated from the Fourier basis discussed above, because it can be considered as the Fourier transform of a Fourier basis function in momentum space (see for instance [19]):

$$\text{Sinc}[(x - x_j)] = \int_{-\infty}^{\infty} e^{-\imath p x} \tilde{\phi}_j(p)dp, \qquad (5.23)$$

where

$$\tilde{\phi}_j(p) = \begin{cases} e^{\imath 2\pi x_j p} & |p| < \pi \\ 0 & |p| > \pi \end{cases} \qquad (5.24)$$

Note, that the optimum grid for interpolation with Sinc functions must also be equi-spaced and centered, $x_j = j \Delta x$, $(j = 0, \pm 1, \pm 2, \ldots, \pm N/2)$. By the properties of the Fourier transform, we see that the discrete version of Eq. 5.23 above will span all the momenta up to the value $p_{max} = \pi/\Delta x$ and therefore Fourier and Sinc pseudospectral methods are completely equivalent in accuracy. The coefficients $b_{k,j}^{(m)}$ which give the approximation to the mth derivative can be obtained by analytically differentiating the Sinc function, $\text{Sinc}[(x - x_j)/\Delta x]$. For the second derivative, which is the case of interest for the kinetic energy term in the Schrödinger equation, the coefficients read:

$$b_{k,j}^{(2)} = \begin{cases} \frac{2(-1)^{j+1}}{j^2 \Delta x^2} & j = \pm 1, \pm 2, \ldots \\ -\frac{\pi^2}{3 \Delta x^2} & j = 0 \end{cases} \qquad (5.25)$$

Note, that these coefficients decay as $\mathcal{O}(1/j^2)$. In common physical applications we want our wavefunction to decay exponentially with j (for instance it can initially be a Gaussian wavepacket), i.e., we want the boundary conditions $u(x \to \infty) = 0$ to be satisfied. Therefore, the derivative sum, Eq. 5.21, will differ from the infinite series by an amount which decreases exponentially with the order N, since, contributions of $u(x_j)$ for large j will be negligible. We can effectively compute the derivative with the accuracy of the full infinite series if N is sufficiently large. The "sufficient" value of N to reach the pseudospectral limit of course depends on the problem we are investigating.

In quantum Molecular Dynamics, the number of grid points (number of terms in the expansion) one should use for a sufficient sampling of the phase space volume, is given by the requirement of "one point per Planck cell" [30], which leads to a relation between the grid spacing and the maximum value of the wavenumber k ($p = k\hbar$) we want to represent:

$$\Delta x = \frac{\pi}{|k_{max}|}. \qquad (5.26)$$

This is precisely the same relation arising in the Sinc or the Fourier PS methods as discussed above. The maximum momentum can be obtained from physical considerations, since, we want in general our wavefunction to be zero at sufficiently distant points of the grid, $\psi(x_{max}) = 0$, and we calculate the potential energy at $x = x_{max}$ and the momentum as $|p_{max}| = \sqrt{2mV(x_{max})}$.

5.1.3 Relations Between FD and PS Methods

That PS and FD approaches must be related can be seen intuitively from the fact that PS methods also provide the exact derivatives of the interpolation polynomial passing through the collocation points. We can be more specific and take the limit $N \to \infty$ of the Lagrange interpolating polynomial. Consider an equi-spaced grid around $x = 0$ with spacing $\Delta x = 1$ extended over $N = 2M + 1$ grid points. The Lagrange fundamental polynomial will be:

$$L_j^M(x) = \frac{(x + M) \cdot (x + M - 1) \cdot \ldots \cdot (x - j + 1) \cdot (x - j - 1) \cdot \ldots \cdot (x - M)}{(j + M) \cdot (j + M - 1) \cdot \ldots \cdot (1) \cdot (-1) \cdot \ldots \cdot (j - M)}.$$

(5.27)

Starting from the central factors, this can be rearranged as the product $\prod_{k=1}^M (1 - (x - j)^2/k^2)$, which in the limit $M \to \infty$ becomes

$$\lim_{M \to \infty} L_j^M(x) = \prod_{k=1}^{\infty} \left(1 - \frac{(x - j)^2}{k^2}\right) = \frac{\sin[\pi(x - j)]}{\pi(x - j)}.$$

(5.28)

Therefore, the infinite order limit of FD gives a PS method with Sinc functions as the expansion basis functions [19]. This has also been noted by Colbert and Miller [12] in the context of a discrete variable representation for the calculation of reaction probabilities.

The infinite order coefficients of a FD approach or equivalently the Sinc DVR expansion coefficients, (Eq. 5.25), decrease only as $O(1/j^2)$ and therefore the approximating series to the derivative of the wavefunction converge slowly if $\psi(x)$ is of the same order of magnitude as the coefficients (i.e., we are above the aliasing limit [39], Eq. 5.26). This means that truncation of the Sinc PS method using less points than the needed from the relation Eq. 5.26 will give very poor results. If we want to improve the convergence of the trigonometric series in order to be able to use less terms (less grid points) in the approximation, we should use an acceleration scheme, which in turn implies to multiply the terms in the series by some acceleration weights. A classical example is the Euler's transformation [36].

A general alternating series is

$$S_M = \sum_{j=0}^{M} a_j z^j,$$

(5.29)

with

$$a_0 = b_{k,0}$$
$$a_j = b_{k,j} \cos (kj \Delta x),$$
(5.30)

where $b_{k,j}$ are the Sinc weights defined in Eq. 5.25 (note that they alternate in sign). Alternating series are ideal candidates for linear acceleration techniques [8, 9]. Boyd has shown that the Mth order Finite Difference approximation is equivalent to the accelerated series [9]:

$$S_M^{FD} = \sum_{j=0}^{M} c_{M,j} a_j z^j,$$
(5.31)

with acceleration weights

$$c_{M,0} = (6/\pi^2) \left\{ \sum_{j=1}^{M} 1/j^2 \right\}$$
$$c_{M,j} = (M!)^2 / [(M - j)!(M + j)!], j = 1, \ldots, M.$$
(5.32)

Since, Sinc functions are the infinite order limit of an equi-spaced FD, the correspondence now is that *periodically repeated* FD stencils will tend to the PS limit of Fourier functions as the number of grid points in the stencil approaches the total number of grid points in one period. Also, because equi-spaced FD can be considered as a robust sum acceleration scheme of a Sinc function series, we expect the same convergence properties of the FD approximation to the Fourier series.

Hence, we have discussed how FD is related to the Sinc-DVR method by taking the limit in the two above mentioned senses:

i) An infinite order limit of centered FD formulae on an equi-spaced grid yields the Discrete Variable Representation (DVR) result when we use as a basis set the Sinc functions ($\text{Sinc}(x) \equiv \sin(\pi x)/\pi x$) [12, 19]. Although, this limit is defined formally as N, the number of grid points used in the approximation, tends to infinity, some theoretical considerations [19] as well as numerical results [22, 23] lead us to expect that the accuracy of the FD approximation is the same to that of the DVR method as we approach the full grid to calculate the FD coefficients.

ii) FD can also be viewed as a sum acceleration method which improves the convergence of the pseudospectral approximation [8]. The rate of convergence is, however, non uniform in the wavenumber, giving very high accuracy for low wavenumbers and poor accuracy for wavenumbers near the aliasing limit [39]. However, this does not cause a severe practical limitation, since, by increasing the number of grid points in the appropriate region we can have an accurate enough representation of the true spectrum in the range of interest. This is one property which makes FD useful as an alternative to the common DVR and other PS methods such as Fast Fourier Transform techniques (FFT) [30].

5.1.3.1 Periodic Non Uniform Grids

For periodic problems trigonometric expansions satisfy all the above stated requirements [30] (the efficiency due to the use of the FFT algorithm), while the Sinc method
is applied in the context of the DVR [12]. As we have discussed it before the two
methods are closely related. For non periodic problems a very successful type of basis
functions is orthogonal polynomials of Jacobi type, with Chebyshev and Legendre
as the most important special cases (see the discussion in Ref. [20]).

Using an orthogonal polynomial basis in Eqs. 5.15 and 5.16, $\phi_k = P_k$, to obtain
the cardinal basis functions we take

$$u_j(x) = w_j \sum_{k=1}^{N} P_k(x_j) P_k(x), \quad j = 1, \ldots, N. \tag{5.33}$$

x_j are the zeros of the polynomial P_{N+1} of degree N. Notice, that in Eq. 5.33 the
summation starts from 1 which corresponds to the constant term in the polynomial.
Hence, P_N denotes a polynomial of degree $N - 1$. The δ-Kronecker property of
the cardinal functions is satisfied by the Christoffel-Darboux theorem for orthogonal
polynomials [32, 47]. To show that the FD formulae are also the limit of orthogonal polynomial expansions in DVR methods, we have to establish the equivalence
between the DVR functions, Eq. 5.33, and the Lagrange fundamental polynomials,
$L_j(x)$, of order $N - 1$.

First we note, that by the definition of Gaussian quadrature $L_j(x)$ satisfy the
orthogonality property:

$$\int_a^b L_i(x) L_j(x) dx = w_j \delta_{ij}, \quad i, j = 1, 2, \ldots, N, \tag{5.34}$$

if they are evaluated at the zeros of some orthogonal polynomial. Then, an expansion
of a function $F(x)$ in orthogonal polynomials can be represented with the series

$$F(x) \approx \sum_{k=1}^{N} F_k P_k(x), \tag{5.35}$$

where the coefficients are

$$F_k = \int_a^b F(x) P_k(x) dx, \tag{5.36}$$

and therefore, $F(x)$ can be defined through the integral

$$F(x) = \int_a^b F(x_j) K_N(x_j, x) dx_j. \tag{5.37}$$

The kernel $K_N(x_j, x)$ is defined by

$$K_N(x_j, x) = \sum_{k=1}^{N} P_k(x_j) P_k(x).$$ (5.38)

On the other hand, expanding $F(x)$ in terms of Lagrange fundamental polynomials, Eq. 5.1, we obtain

$$F(x) = \sum_{k=1}^{N} F(x_k) L_k(x) = \int_a^b F(x_j) w_j^{-1} \sum_{k=1}^{N} L_k(x_j) L_k(x) dx_j,$$ (5.39)

taking into account the orthogonality relation, Eq. 5.34. Since, the kernels in (5.37) and (5.39) must coincide, we have that

$$L_j(x) = w_j \sum_{k=1}^{N} P_k(x_j) P_k(x),$$ (5.40)

using the δ-Kronecker property of $L_j(x)$.

Hence, we have shown that when we use the N roots of the P_{N+1} polynomial as interpolating grid points the Lagrange fundamental polynomials are the cardinal functions that correspond to the orthogonal polynomials $P_k(x)$, $k = 1, \ldots, N$.

5.1.4 Remarks

The current interest in Finite Difference methods is fully justified when solutions of the Schrödinger equation are required for multidimensional systems such as polyatomic molecules. The present most popular methods employed in quantum molecular dynamics are the Fast Fourier Transform and the Discrete Variable Representation techniques. FFT generally uses hypercubic grid domains which result in wasted configuration space sampling. A large number of the selected configuration points correspond to high potential energy values, which do not contribute to the eigenstates that we are seeking. Global DVR methods allow us to choose easily the configuration points which are relevant to the states we want to calculate, but still, we must employ in each dimension all grid points. Local methods such as FD have the advantages of DVR but also produce matrices with less non zero matrix elements provided that the PS accuracy is achieved at lower order than the high order limit.

There are some other benefits for FD with respect to global pseudospectral methods. Convergence can be examined not only by increasing the number of grid points but also by varying in a systematic way the order of approximation of the derivatives. Finite Difference methods may incorporate several boundary conditions and choose the grid points without necessarily relying on specific basis functions. The topography of the multidimensional molecular potential functions is usually complex. The ability of using non equi-spaced grids is as important as keeping the grid points in accordance to the chosen energy interval. The computer codes for a FD representation of the Hamiltonian can be parallelized relatively easily, since, the basic operation is the multiplication a sparse matrix by a vector. Parallelization is an obligatory task when we deal with systems of more than three degrees of freedom and we look for highly excited states.

Sinc-DVR methods are appropriate for radial variables where the wavefunction must vanish at the edge of the grid ($\psi(R) = 0$ for $a \geq R \geq b$). The FD weights required in approximating the derivatives of the wavefunction close to the borders of the grid can be calculated for this boundary condition by extending the grid intervals with fictitious points. Another type of radial coordinates frequently encountered in molecular dynamics are those which can not be extended with fictitious points. Such a variable is the distance of an atom from the center of mass of a diatomic molecule in Jacobi coordinates which may start from zero for linear configuration. In this case it is necessary to employ one-sided FD formulae.

In summary, we find the following advantages of FD approach in solving the Schrödinger equation:

1. Finite Differences allow a systematic search of the convergence properties with respect to the number of grid points as well as the order of approximation of the derivatives.
2. We can use a large number of grid points for better representing the wave function and save computer time and memory by employing low order approximations.
3. FD with stencils the total number of grid points are equivalent to the most common PS methods (Sinc, Fourier, Chebyshev, Legendre).
4. Truncated PS methods are generally bad approximations, whereas Finite Differences show smooth convergence behaviour by increasing the order.
5. There is flexibility in choosing the grid points without necessarily any dependence on specific basis functions.
6. Algorithms for a fast generation of the weights in the FD approximations of the derivatives by recursion relations are available.
7. The computer codes can be parallelized.

5.2 Shooting Methods

Locating periodic orbits may be seen as a 2-point boundary value problem. The boundary conditions are the equations of closing the trajectory in phase space. There are two classes of numerical methods for solving in general 2-point boundary value problems [39]. The *shooting methods* involve those in which the 2-point boundary value problem is converted to an *initial value* one. Choosing initial values for the trajectory we integrate the equations of motion and check the non closure of the trajectory. By varying the initial conditions or some free parameters we successively approach the trajectory which satisfies the boundary conditions.

The second class is referred to the *relaxation methods*. In these the differential equations are replaced with difference equations by discretizing the variables. Then, starting with an approximate solution we try to bring it into successively closer agreement with the finite difference equations, and the boundary conditions.

Both classes of methods have been applied to locate periodic orbits in molecular systems. The shooting methods, also known as Newton methods, are the most popular [18]. There are several variants of it resulted from fixing the total energy or the period of the periodic orbit, using or not a Poincaré surface of section, and using analytical second derivatives of the Hamiltonian or numerically estimating the gradient in the Newton-Raphson method by integrating neighboring trajectories. The *Monodromy Method* of Baranger and coworkers [1, 6, 14] is a technique which is classified in the relaxation methods.

An extension of the shooting techniques which tries to incorporate the benefits of the relaxation technique is the *multishooting method* [15, 16, 28, 40, 41, 45]. In this case the 1-point initial value problem is converted to $(m - 1)$ initial value ones by choosing m nodes in the independent variable. We do not take a finite difference representation of the equations of motion but instead, we integrate $(m-1)$ trajectories and by varying their $(m - 1)$ initial conditions we approach to a smooth trajectory which satisfies the boundary values.

5.2.1 The 2-Point Boundary Value Problem

Taking q^i, $(i = 1, \ldots, n)$ generalized coordinates and p_i, $(i = 1, \ldots, n)$ conjugate momenta as the components of the vector x,

$$x = (q^1, \ldots, q^n, p_1, \ldots, p_n)^T, \tag{5.41}$$

where (T) denotes the transpose of the $(2n)$D row vector. The equations of motion written in components are,

$$\dot{x}^\mu \equiv \frac{dx^\mu}{dt} = \sum_{\nu=1}^{2n} J^{\mu\nu} \frac{\partial H}{\partial x^\nu} \equiv \sum_{\nu=1}^{2n} J^{\mu\nu} \partial_\nu H, \quad \mu = 1, \ldots, 2n. \tag{5.42}$$

H is the Hamiltonian and J the symplectic matrix (Eq. 2.49).

The equilibrium points are defined by requiring $\dot{x} = 0$, or

$$\partial_\nu H(x) = 0, \quad \nu = 1, \ldots, 2n. \tag{5.43}$$

If $x(0)$ denotes the initial conditions of a trajectory at time $t_1 = 0$, then this trajectory is periodic if it returns to its initial point in phase space after the time

$t_2 = T$ (period). Hence, the Hamiltonian flow Φ_t satisfies the condition

$$\Phi_T[x(0)] = x(0). \tag{5.44}$$

Thus, to find periodic solutions it is necessary to solve the above nonlinear equations.

5.2.2 The Initial Value Problem

The above boundary value problem is converted to an *initial value problem* by considering the initial values of the coordinates and momenta s

$$x(0) = s, \tag{5.45}$$

as independent variables in the nonlinear functions

$$B(s) = x(T; s) - s. \tag{5.46}$$

We denote the roots of Eqs. 5.46 as s_*, i.e.,

$$B(s_*) = 0. \tag{5.47}$$

Thus, if s is a nearby value to the solution s_* we can compute the functions $B(s)$ by integrating Hamilton's equations for the period T. By appropriately modifying the initial values s we hope to converge to the solution, that is $s \rightarrow s_*$ and $B \rightarrow 0$.

5.2.3 The Newton-Raphson Iterative Method

The common procedure to find the roots of Eq. 5.47 is the Newton-Raphson method. This is an iterative scheme and at each iteration, k, we update the initial conditions of the orbit

$$s_{k+1} = s_k + \Delta s_k. \tag{5.48}$$

The corrections Δs_k are obtained by expanding Eq. 5.46 in a Taylor series up to the first order

$$B(s_{k+1}) \approx B(s_k) + \frac{\partial B}{\partial s_k} \Delta s_k = 0,$$

$$B(s_k) + \left[\frac{\partial x_k(T; s_k)}{\partial s_k} - I_{2n} \right] \Delta s_k = 0, \tag{5.49}$$

where at the kth iteration

$$B(s_k) = x_k(T; s_k) - s_k. \tag{5.50}$$

The Jacobian matrix

$$Z_k(T) = \frac{\partial x_k(T; s_k)}{\partial s_k}, \tag{5.51}$$

is the Fundamental Matrix which is evaluated by integrating the variational equations (see Sect. 3.2.1 and Eq. 3.25). Thus, to complete the kth iteration in the Newton-Raphson method we first integrate for time T the differential equations

$$\dot{x}_k(t) = J \partial H[x_k(t)]$$
$$\dot{Z}_k(t) = A_k(t) Z_k(t), \tag{5.52}$$

with initial conditions

$$x_k(0) = s_k$$
$$Z_k(0) = I_{2n}. \tag{5.53}$$

Then, we solve the linear algebraic equations

$$[Z_k(T) - I_{2n}] \Delta s_k = -B(s_k), \tag{5.54}$$

in order to find the initial conditions for the $(k+1)$th iteration (Eq. 5.48).

5.2.4 The Underrelaxed Newton-Raphson Method

Quite often the Newton-Raphson method diverges, although, when it converges it does that quadratically. Sometimes problems of divergencies are cured by scaling the corrections with a parameter λ_k

$$s_{k+1} = s_k + \lambda_k \Delta s_k, \tag{5.55}$$

where $0 \leq \lambda_k \leq 1$, and $\lambda_k \to 1$ as $s_k \to s_*$. Several schemes for selecting λ_k have been proposed [15]. A simple one is

$$\lambda_k = \frac{\lambda_{min}}{max(\lambda_{min}, ||\Delta s_k||)}. \tag{5.56}$$

λ_{min} is an input minimum value for the parameter, and $||\ ||$ denotes the norm of the vector (Euclidean). For convergence criteria we use the norms

$$||B(s_k)|| = \left[\sum_{i=1}^{2n} B_i(s_k)^2 \right]^{1/2} < d_1, \qquad (5.57)$$

and

$$||\Delta s_k|| = \left[\sum_{i=1}^{2n} \Delta s_{i(k)}^2 \right]^{1/2} < d_2. \qquad (5.58)$$

The linear system of Eq. 5.54, Δs_k, may be solved by several algorithms; (i) LU-decomposition methods, (ii) SVD-Singular Value Decomposition, and (iii) Iterative methods such as the conjugate gradient, variable metric, and quasi-Newton methods [39].

5.2.5 The Multiple Shooting Method

The idea of multishooting is to combine shooting and relaxation techniques. Let us assume that we divide the period T in $(m - 1)$ time intervals, while first for convenience we introduce a new scaled time $\tau = t/T$, $\quad (0 \le \tau \le 1)$,

$$0 = \tau_1 < \tau_2 < \cdots < \tau_{m-1} < \tau_m = 1. \qquad (5.59)$$

Thus, for the simple shooting method $m = 2$.

From now on we drop the index for the iterations k, and we use the index j to denote the nodes in the periodic orbit. If the initial conditions of the trajectory at each node j is s_j at time τ_j, and the final value of the trajectory at time τ_{j+1} is denoted by $x(\tau_{j+1}; s_j)$, then $(m - 2)$ *continuity conditions* should be satisfied

$$C_j(s_j, s_{j+1}) = x(\tau_{j+1}; s_j) - s_{j+1} = 0, \quad j = 1, 2, \ldots, m - 2, \qquad (5.60)$$

together with the boundary conditions

$$B(s_{m-1}, s_1) = x(\tau_m; s_{m-1}) - s_1 = 0. \qquad (5.61)$$

Now, we have to solve $(m - 1)$ initial value problems, and for that we follow the linearized Newton-Raphson method of the previous sections

$$C_j(s_j, s_{j+1}) + \frac{\partial C}{\partial s_j} \Delta s_j + \frac{\partial C}{\partial s_{j+1}} \Delta s_{j+1} = 0, \qquad (5.62)$$

which become

$$C_j(s_j, s_{j+1}) + Z_j(\tau_{j+1})\Delta s_j - \Delta s_{j+1} = 0, \quad 1 \le j \le m - 2. \tag{5.63}$$

Using the boundary conditions (Eq. 5.61) we get

$$B(s_{m-1}, s_1) + Z_{m-1}(\tau_m)\Delta s_{m-1} - \Delta s_1 = 0, \tag{5.64}$$

where,

$$Z_j(\tau_{j+1}) = \frac{\partial x(\tau_{j+1}; s_j)}{\partial s_j}. \tag{5.65}$$

Eqs. 5.63 and 5.64 are written in a matrix form of dimension $2n(m-1) \times 2n(m-1)$

$$\begin{bmatrix} Z_1 & -I_{2n} & 0 & \cdots & 0 & 0 \\ 0 & Z_2 & -I_{2n} & \cdots & 0 & 0 \\ \cdots & \cdots & \cdots & \cdots & \cdots & \cdots \\ 0 & 0 & 0 & \cdots & Z_{m-2} & -I_{2n} \\ -I_{2n} & 0 & 0 & \cdots & 0 & Z_{m-1} \end{bmatrix} \begin{bmatrix} \Delta s_1 \\ \Delta s_2 \\ \cdots \\ \Delta s_{m-2} \\ \Delta s_{m-1} \end{bmatrix} = - \begin{bmatrix} C_1 \\ C_2 \\ \cdots \\ C_{m-2} \\ B \end{bmatrix} \tag{5.66}$$

The above system of linear equations is solved by invoking the so called *condensing algorithm* [28]

$$\Delta s_1 = -E^{-1}u, \tag{5.67}$$

$$\Delta s_{j+1} = Z_j \Delta s_j + C_j, \quad j = 1, 2, \ldots, m - 2, \tag{5.68}$$

where

$$E = Z_{m-1}Z_{m-2}\cdots Z_2 Z_1 - I_{2n},$$
$$u = B + Z_{m-1}(C_{m-2} + Z_{m-2}(C_{m-3} + Z_{m-3}(C_{m-4} + \cdots + Z_2 C_1)\cdots)).$$

5.2.6 Implementation

The $(m-1)$ fundamental matrices required in the multishooting method may be evaluated either from numerically obtained derivatives or analytically. The first requires the integration of $2n(m-1)$ neighboring trajectories, and the derivatives are then computed by finite differences. In the case that the analytic second derivatives of the Hamiltonian are available, we integrate Hamilton's and the variational equations together, Eq. 5.52, with initial conditions the Eq. 5.53. After converging to the periodic orbit we can have an estimate of the Monodromy Matrix from the product of matrices.

$$M = Z_{m-1}Z_{m-2}\cdots Z_2Z_1. \tag{5.69}$$

Sometimes it is desirable to bring all periodic orbits on a common Poincaré surface of section. The Hénon method [26] is not suitable for highly unstable systems. Then, it is more convenient to increase the boundary conditions by fixing one coordinate (momentum), i.e.,

$$x_l - \xi = 0, \tag{5.70}$$

and to consider, that the period of the periodic orbit satisfies the trivial differential equation [41]

$$\dot{T} = 0. \tag{5.71}$$

Thus, a $(2n + 1)$ dimensional boundary value problem must be solved.

In the continuation of a family of periodic orbits we found it useful to vary the period as a parameter, and for that we use predictor-corrector algorithms with trivial or secant predictors [2].

5.3 Computer Software

5.3.1 GridTDSE

We have developed a parallel FORTRAN 95 code (GridTDSE) [46] for obtaining the solutions of the time dependent Schrödinger equation and the spectrum of the Hamiltonian by filter-diagonalization [49] and Lanczos methods [31].

5.3.1.1 Discretization of the Wavefunction

In the implementation of the code the first step is discretization of the wavefunction in the coordinate space. The Finite Difference method is introduced as a local approximation of a certain function $u(x)$, which is represented at the grid points with an ensemble of n_s ($n_s = 2s + 1$ with s an integer) Lagrange interpolation polynomials. This yields the approximation formula for the derivatives:

$$\frac{\mathrm{d}^m u(x)}{\mathrm{d}x^m}\bigg|_{x=x_k} \approx \sum_{j=1}^{n_s} b_{k,j}^m u(x_j), \tag{5.72}$$

where the $b_{k,j}^m$ coefficients are computed by using Fornberg's algorithm [19]. For the case $m = 0$, Eq. 5.72 turns into a $(n_s - 1)$-order polynomial interpolation formula of $u(x)$.

Adopting a center of mass Cartesian coordinate system the wavefunction is represented by a vector whose $\prod_{i=1}^{3N-3} n_i$ elements are the values of the wavefunction at each grid point, with n_i points per dimension (i). Equation 5.72 is then equivalent to the action of a matrix (differentiation matrix) on this vector. The n_s-point truncation (called *stencil*) in the series of Eq. 5.72 means that information from the $\pm s$ neighbouring points plus the value of the function at the specific grid point are needed for the evaluation.[1] The accuracy of the approximation improves very fast while the number of points in the stencil n_s is increased.

5.3.1.2 Evaluation of $\hat{H}\psi_J$

The evaluation of the action of the Hamiltonian operator over the projected into the proper angular momentum subspace initial wavepacket requires to compute $\hat{H}\psi_J$ (Eq. 4.88). This is needed in the time propagation step as we shall see in the next subsection, but also in extracting the eigenenergies and eigenfunctions by iterative methods such as filter-diagonalization [11, 33–35, 49] and Lanczos [10, 25, 31]. In a discretized scheme and a FD approximation of the Hamiltonian, this operation turns into a sparse matrix-vector multiplication, which can be calculated with the appropriate computational algorithms for linear systems.

The Hamiltonian matrix encompasses the potential energy and the kinetic energy operators. The first one refers to a local property of the system, and thus it is represented by a diagonal matrix. For the second one, the Laplacian is evaluated using Eq. 5.72. The sparsity of the Hamiltonian depends on the length of the stencil ($n_s = 2s + 1$) employed in the calculation of the Laplacian, which is a matrix with $(3N-3)(n_s-1)+1^2$ non zero matrix elements per row. In the present version of the computer code, we use matrix-vector multiplication subroutines from the Portable Extensible Toolkit for Scientific Computation (PETSc) [3–5].

5.3.1.3 Time Evolution of the Wavepacket

The final step of the algorithm involves the propagation of the wavepacket in time. There are several efficient ways to calculate the evolution operator $U(t, \hat{H})$ by means of a polynomial expansion [29, 50], two of them have been implemented in our code. The Second Order Difference (SOD) [30] method has the main advantage of the simplicity of the algorithm. The time propagation is performed using the recursive formula:

$$\psi(t+\Delta) = \psi(t-\Delta) - \frac{2i\Delta}{\hbar}\hat{H}\psi(t), \tag{5.73}$$

---

[1] We assume a center difference scheme to compute the derivatives, but more general schemes can be adopted [19].

[2] We assume stencils with the same number of points in each dimension.

where Δ is the time step. Wavefunctions at two previous times are needed; therefore, apart from the known initial wavefunction, in order to get the next wavefunction we apply at the beginning of the propagation an Euler scheme [39] to advance the wavepacket by one time step. Although the wavefunction is propagated conserving the norm and the total energy, the approximation is only correct up to the second order ($O(\Delta^2)$), requiring small time steps in the propagation. Higher order polynomial approximations of the evolution operator have been published [51] and implemented in our code.

The second method is based on a Chebyshev expansion of the propagator, and it has been explained in detail by Tal-Ezer and Kosloff in [48]. It can be viewed as an interpolation in the energy space, where the propagator $U(t, \hat{H})$ is expanded on a basis of $M + 1$ Chebyshev polynomials T_k [39] which are functions of a normalised Hamiltonian $\tilde{H}$

$$
\psi(t + \Delta) \approx \sum_{k=0}^{M} a_k T_k \left(\frac{-\imath \tilde{H} \Delta}{\hbar} \right) \psi(t). \tag{5.74}
$$

The convergence in the series is ensured by the rapid exponential decay of the Bessel functions, resulting in a highly accurate method. Unlike the SOD method, Chebyshev scheme is not unitary by construction, and neither the norm nor the energy are guaranteed to be conserved. The main advantage of this method is that calculations with long time step, Δ, can be accurately carried out by increasing the order in the expansion. However, this scheme implies that any intermediate time information is lost. In order to follow the time evolution during the process, the scheme has to be repeated many times, which yields a linear propagation of error, ε, in first-order approximation.

Finally, as was mentioned before (Sect. 4.2), properties of the molecule can be calculated from the autocorrelation function $C(t)$:

$$
C(t) = < \psi(x, 0) | \psi(x, t) > = \int \psi^*(x, 0) \psi(x, t) \mathrm{d}x, \tag{5.75}
$$

and its Fourier transform $I(E)$, Eq. 4.45.

For infinite integration time and bound systems, the quantity $I(E)$ consists of an ensemble of delta functions that are placed at the eigenvalues of the molecule. Thus, the energy resolution of this representation depends on the characteristics of the time integration scheme. Increasing the resolution of the spectrum, and thus the accuracy of the energy levels, implies a longer time integration.

For multidimensional systems, evaluating $\hat{H}\psi$ becomes the most cumbersome task of the program. The parallelization of the code takes place at each matrix-vector operation. For the moment, there are no general parallelised codes for Quantum Molecular Dynamics. In our case, parallelization is implemented by employing several subroutines of PETSc, which is a set of libraries that manipulate a particular family of objects, among others vectors and matrices. PETSc is especially efficient for carrying out parallel vector-vector and matrix-vector operations. PETSc bases its parallelization on the Message Passing Interface (MPI)[3] for communication among processors.

Internally, the matrix is stored in a Compressed Row Storage (CRS) format, appropriate for sparse matrices. This is implemented by three vectors: One for storing the non zero elements of the matrix, one integer vector for the corresponding column indexes and another integer vector for the locations that start a row. In the FD scheme, the dimensions of these vectors would be $[N_{dim}(n_s - 1) + 1] n_g^{N_{dim}}$ for the first two and $n_g^{N_{dim}} + 1$ for the last integer vector, where $n_g^{N_{dim}}$ would be the dimension of the wavefunction and N_{dim} the number of Cartesian coordinates [46].

A convenient rectangular coordinate space grid is usually employed. However, the spherical symmetry of the potential reduces the suitability of the rectangular grid. In fact, we know that the relative volume of the hypersphere to the hypercube decreases rapidly with the dimensionality as

$$\pi^{N_{dim}/2} / \left(2^{N_{dim}} \Gamma(N_{dim}/2 + 1) \right).$$

$\Gamma(N_{dim}/2 + 1)$ denotes the gamma function. A very convenient way to increase the efficiency of the grid and to impose correct boundary conditions is to use a cut-off value (V_c) for the potential, neglecting the contribution of those points where the potential is over the pre-defined threshold, $V(x) > V_c$, as there the wavefunction is approximately equal to zero. As is expected, the effect of introducing the cut-off is more pronounced in the high-dimensional case.

5.3.2 POMULT

POMULT is a FORTRAN 90 code for locating Periodic Orbits and Equilibrium Points in Hamiltonian systems based on 2-point boundary value solvers which use multiple shooting algorithms [17]. The code has mainly been developed for locating periodic orbits in molecular Hamiltonian systems with many degrees of freedom and it utilizes a damped Newton-Raphson method and a secant method. POMULT provides routines for a general analysis of a dynamical system such as fast Fourier transform of the trajectories, Poincaré surfaces of sections, maximal Lyapunov exponents and evaluation of the classical autocorrelation functions and power spectra.

We have adopted to POMULT the Molecular Mechanics suite of programs, TINKER [38] in order to calculate empirical potential functions for polyatomic molecules together with their first (for solving Hamilton's equations) and second (for solving the variational equations) derivatives analytically.

---

[3] We have used OPEN-MPI to compile PETSc (http://www.open-mpi.org/).

5.3.3 Normal Forms

To construct tori, NHIM and (un)stable manifolds we need to transform a global Hamiltonian in physical coordinates to a local one in normal forms around a particular equilibrium point. Software for computing the classical and semi-classical normal forms in the neighbourhood of equilibria of n-degrees-of-freedom Hamiltonian systems has been developed at the School of Mathematics of the University of Bristol by Wiggins and collaborators [13]. The code is written in C++ and PYTHON.

The program provides routines for forward and backward transformations between physical and normal form coordinates (Sect. A.10). This is important when we select initial conditions of trajectories lying on invariant phase space structures and for integrating them in the physical coordinates of the system.

5.3.4 AUTO_DERIV

AUTO_DERIV [42, 43] is a module comprised of a set of FORTRAN 90 procedures which can be used to calculate the first and second partial derivatives of any continuous function with many independent variables. The function should be expressed as one or more FORTRAN 90 or FORTRAN 77 procedures. A new type of variables is defined and the overloading mechanism of functions and operators provided by the FORTRAN 90 language is extensively used to define the differentiation rules.

References

1. Aguiar MAM, Malta CP, Baranger M, Davies KTR (1987) Bifurcations of periodic trajectories in non-integrable Hamiltonian systems with two degrees of freedom: numerical and analytical results. Annals Phys 180:167
2. Allqower EL, Georg K (1990) Numerical continuation methods., Springer series in computational mathematics, Springer, Berlin
3. Balay S, Gropp WD, McInnes LC, Smith BF (1997) Efficient management of parallelism in object oriented numerical software libraries. In: Arge E, Bruaset AM, Langtangen HP (eds) Modern software tools in scientific computing, Birkhäuser Press, pp 163–202
4. Balay S, Buschelman K, Gropp WD, Kaushik D, Knepley MG, McInnes LC, Smith BF, Zhang H (2001) PETSc Web page. http://www.mcs.anl.gov/petsc
5. Balay S, Buschelman K, Eijkhout V, Gropp WD, Kaushik D, Knepley MG, McInnes LC, Smith BF, Zhang H (2004) PETSc users manual. Technical Report ANL-95/11— Revision 2.1.5, Argonne national laboratory.
6. Baranger M, Davies KTR, Mahoney JH (1988) The calculation of periodic trajectories. Annals Phys 186:95–110
7. Baye D, Heenen PH (1986) Generalised meshes for quantum mechanical problems. J Phys A: Math Gen 19:2041–2059
8. Boyd JP (1992) A fast algorithm for Chebyshev, Fourier, and Sinc interpolation onto an irregular grid. J Comp Phys 103:243–257

9. Boyd JP (1994) Sum-accelerated pseudospectral methods: finite differences and sech-weighted differences. Comp Methods Appl Mech Engrg 116:1–11

10. Carrington T Jr (2004) Methods for calculating vibrational energy levels. Can J Chem 82:900–914

11. Chen R, Guo H (1996) A general and efficient filter-diagonalization method without time propagation. J Chem Phys 105(4):1311–1317

12. Colbert DT, Miller WH (1992) A novel discrete variable representation for quantum mechanical reactive scattering via the S-matrix Kohn method. J Chem Phys 96:1982–1991

13. Collins P, Burbanks A, Wiggins S, Waalkens H, Schubert R (2008) Background and documentation of software for computing Hamiltonian normal forms. School of mathematics, University of Bristol, University Walk, Bristol BS8 1TW, 1st edn

14. Davies KTR, Huston TE, Baranger M (1992) Calculations of periodic trajectories for the Hénon-Heiles Hamiltonian using the monodromy method. Chaos 2:215–224

15. Deuflhard P (1974) A modified Newton method for the solution of ill-conditioned systems of nonlinear equations with application to multiple shooting. Numer Math 22:189–315

16. Deuflhard P (1979) A stepsize control for continuation methods and its special application to multiple shooting techniques. Numer Math 33:115–146

17. Farantos SC (1998) POMULT: a program for computing periodic orbits in Hamiltonian systems based on multiple shooting algorithms. Comp Phys Comm 108:240–258

18. Feudel U, Jansen W (1992) CANDYS/QA—a software system for the qualitative analysis of nonlinear dynamical systems. Int J Bifurc and Chaos 2:773–794

19. Fornberg B (1998) A practical guide to pseudospectral methods. Cambridge monographs on applied and computational mathematics, Cambridge University Press, Cambridge

20. Fornberg B, Sloan DM (1994) A review of pseudospectral methods for solving partial differential equations. Acta Numerica 3:203–267

21. Gradshteyn IS, Ryzhik IM (1994) Table of integrals, series and products. Academic Press, New York

22. Guantes R, Farantos SC (1999) High order finite difference algorithms for solving the Schrödinger equation in molecular dynamics. J Chem Phys 111:10,827–10,835

23. Guantes R, Farantos SC (2000) High order finite difference algorithms for solving the Schrödinger equation in molecular dynamics II: periodic variables. J Chem Phys 113:10,429–10,437

24. Guantes R, Nezis A, Farantos SC (1999) Periodic orbit—quantum mechanical investigation of the inversion mechanism of Ar_3. J Chem Phys 111:10,836–10,842

25. Guo H (2007) Recursive solutions to large eigenproblems in molecular spectroscopy and reaction dynamics. Rev Comput Chem 25:285–347

26. Hénon M (1982) On the numerical computation of Poincaré maps. Physica D 5:412–414

27. Karabulut H, Sibert EL III (1997) Trigonometric discrete variable representations. J Phys B: At Mol Opt Phys 30:L513–L516

28. Keller HB (1976) Regional Conf Ser in Appl Math Numerical solution of two point boundary value problems SIAM. 24:61

29. Kosloff R (1994) Propagation methods for quantum molecular dynamics. Annu Rev Phys Chem 45:145–178

30. Kosloff R (1996) Quantum molecular dynamics on grids. In: Zhang JZH Wyatt RE (eds) Marcel Dekker Inc, Dynamics of molecules and chemical reactions, pp 185–230

31. Lanczos C (1950) An iteration method for the solution of the eigenvalue problem of linear differential and integral operators. J Res Natl Bur Stand 45:255–282

32. Light JC, Hamilton IP, Lill JV (1985) Generalized discrete variable approximation in quantum mechanics. J Chem Phys 82:1400–1409

33. Mandelshtam VA, Taylor HS (1995) Spectral projection approach to the quantum scattering calculations. J Chem Phys 102(19):7390–7399

34. Mandelshtam VA, Taylor HS (1997) The quantum resonance spectrum of the H_3^+ molecular ion for J=0. An accurate calculation using filter-diagonalization. J Chem Soc, Faraday Trans 93:847–860

35. Mandelshtam VA, Grozdanov TP, Taylor HS (1995) Bound states and resonances of the hydroperoxyl radical HO_2: an accurate quantum mechanical calculation using filter-diagonalization. J Chem Phys 103(23):10,074–10,084
36. Mathews J, Walker RL (1970) Mathematical methods of physics. Addison-Wesley, Reading
37. Muckerman JT (1990) Some useful discrete variable representations for problems in time-dependent and time-independent quantum mechanics. Chem Phys Lett 173:200–205
38. Ponder JW (2014) Tinker molecular modelling http://dasher.wustl.edu/tinker/
39. Press WH, Teukolsky SA, Vetterling WT, Flannery BP (1992) Numerical recipies. Cambridge University Press, Cambridge
40. Reithmeier E (1991) Periodic solutions of nonlinear dynamical systems. Lecture notes in mathematics, Springer
41. Seydel R (1988) From equilibrium to chaos: practical bifurcation and stability analysis. Elsevier, New York
42. Stamatiadis S, Farantos SC (2010) AUTO_DERIV: tool for automatic differentiation of a FORTRAN code (New Version). Comp Phys Comm 181(10):1818–1819
43. Stamatiadis S, Prosmiti R, Farantos SC (2000) AUTO_DERIV: tool for automatic differentiation of a FORTRAN code. Comp Phys Comm 127:343–355
44. Stenger F (1981) Numerical methods based on Whittaker cardinal, or Sinc functions. SIAM Rev 23:165–224
45. Stoer J, Bulirsch R (1980) Introduction to numerical analysis. Springer, New York
46. Suarez J, Farantos SC, Stamatiadis S, Lathouwers L (2009) A method for solving the molecular Schrödinger equation in Cartesian coordinates via angular momentum projection operators. Comp Phys Comm 180:2025–2033
47. Szegö G (1948) Orthogonal polynomials. Am Math Soc Colloq Publ 23:42
48. Tal-Ezer H, Kosloff R (1984) An accurate and efficient scheme for propagating the time dependent Schrödinger equation. J Chem Phys 81(9):3967–3971
49. Wall MR, Neuhauser D (1995) Extraction, through filter-diagonalization, of general quantum eigenvalues or classical normal mode frequencies from a small number of residues or a short-time segment of a signal. I. Theory and application to a quantum-dynamics model. J Chem Phys 102(20):8011–8022
50. Zhang DH, Zhang JZH (1996) Time-dependent quantum dynamics for gas-phase and gas-surface reactions. In: Zhang JZH Wyatt RE (eds) Dynamics of molecules and chemical reactions, Marcel Dekker Inc, pp 231–276
51. Zhu W, Zhao X, Tang Y (1996) Numerical methods with high order of accuracy applied in the quantum system. J Chem Phys 104(6):2275–2286

Chapter 6
Applications

In this chapter we review a few results from the application of the nonlinear Hamiltonian mechanical methodologies developed in the previous chapters in order to demonstrate their power in unveiling the intriguing indeed behaviour of highly excited molecules. Nonlinear mechanics have been applied to molecular dynamics for several type of molecules, in electronic ground and excited states, and a recent review of this work has been published [6].

The theoretical treatment of molecular spectroscopy and reaction dynamics involves the construction of the potential energy surfaces pertinent to the energy range we study, and then quantum and/or classical mechanical calculations to extract spectra and rate constants. Progress in the experimental techniques has imposed considerable requirements to the theory. Although, methodologies in calculating ab initio PES for polyatomic molecules have made significant progress, accurate quantum dynamical calculations are mainly restricted upto four-five atom molecules with six-nine internal degrees of freedom because of computer limitations. Nevertheless, for polyatomic molecules even if the quantum dynamical calculations were possible, the interpretation of the experimental and numerical results would require one to extract the physics by employing low dimensional models. For highly excited or reacting molecules we have already argued that the utilization of nonlinear mechanics is unavoidable.

Molecular spectroscopy has seen significant advances in both frequency and time domain in the last decades [40]. Techniques such as stimulated emission pumping (SEP), dispersed fluorescence (DF), and high resolution Fourier transform and laser spectroscopy have contributed to the detailed study of small polyatomic molecules [1]. Laser femtosecond spectroscopy and molecular beams [12, 40] have allowed spectroscopists and dynamicists to study isolated molecules and to follow a chemical reaction in real time, where bonds are broken and new ones are formed. Furthermore, spectroscopic methods for studying structural and dynamic properties of complex molecules such as multiple dimensional NMR and optical spectroscopy utilizing multiple ultrafast coherent laser pulses have allowed the study of protein structure and dynamics and femtosecond solvation dynamic [11, 38, 39].

© The Author(s) 2014   103
S.C. Farantos, *Nonlinear Hamiltonian Mechanics Applied to Molecular Dynamics*,
SpringerBriefs in Electrical and Magnetic Properties of Atoms, Molecules, and Clusters,
DOI 10.1007/978-3-319-09988-0_6

A result of the nonlinear mechanical behaviour of a dynamical system at high energies is the coexistence of ordered motions and chaos, as well as the genesis of new type of motions via bifurcation phenomena. As a matter of fact, the progress of nonlinear mechanics forces us to reexamine the mechanisms of the breaking and/or forming a single chemical bond as it happens in elementary chemical reactions. New assignment schemes which allow the classification of quantum states in a meaningful and useful way are required and such novel methods have indeed been developed thanks to the theory of periodic orbits (PO), their bifurcations and the semiclassical quantization theories. The concept of the *transition state* in chemical reaction theories has found a new and more accurate interpretation formulated on Normally Hyperbolic Invariant Manifolds (NHIM) and their stable/unstable manifolds [32].

Molecules are complex systems and as experiments provide more dynamical details all theories, quantum, classical and semiclassical, must be combined to interpret the results. Nonlinear mechanics and their stratified methodologies are of paramount importance to successfully accomplish the endeavour. In this chapter we present a few examples to demonstrate how by applying nonlinear Hamiltonian mechanics, after the construction of the pertinent PES, reveal the dynamics of excited molecules and its reactivity. Ab Initio PES can be used in dynamical calculations either by fitting the calculated energies to analytical functions, including interpolation polynomial like cubic splines and Lagrange polynomials, or by doing calculations on the fly, i.e., by computing first and second derivatives simultaneously to the energies with an ab initio method. However, for large molecules the use of empirical analytical potential functions the parameters of which are fitted to experimental or theoretical data is at present ineluctable because of computer limitations.

In the past years, we have applied the methods of locating PO developed for small molecules to biological molecules, such as peptides described with empirical potential functions. Combining POMULT with TINKER molecular dynamics software to construct the PES, we have studied alanine dipeptide [4] as a prototype system, and we have shown how one can systematically trace regions in phase space where the trajectories stay localized in specific vibrational modes of a conformation or of a transition state. With continuation techniques we obtain families of periodic orbits for an extended energy range and we find elementary bifurcations such as Hamiltonian center-saddle (CS) and Hopf like [5, 14]. In a similar fashion, we have studied the active site of cytochrome *c* oxidase [2], the enzyme which catalyses oxygen molecule to water and contributes in the production of the energy in the cells. In the following sections we present a few results from this work.

6.1 Small Polyatomic Molecules and Ab Initio Potentials

By studying periodic orbits in a parameter space we discover their bifurcations and possible localized eigenstates along them. As it was discussed in Chap. 3, periodic orbits which emerge from center-saddle bifurcations appear abruptly at some critical values of the energy, usually in one pair (two branches), and change drastically the phase space around them. They penetrate in regions of nuclear phase space which the normal mode motions cannot reach. Center-saddle bifurcations are of generic type, i.e., they are robust and remain for small (perturbative) changes of the potential function [21, 22].

To the best of our knowledge, phosphaethyne (HCP) was the first molecule where CS bifurcations were identified spectroscopically (Ref. [13] and references therein). In this study, complementary experimental and theoretical examinations showed the evolution of specific spectral patterns from the bottom of the potential well up to excitation energies of approximately 25,000 cm^{-1}, where large amplitude, isomerization type motion from HCP to CPH is prominent. Distinct structural and dynamical changes, caused by an abrupt transformation from essentially HC bonding to mainly PH bonding, set in around 13,000 cm^{-1}. They reflect center-saddle bifurcations of periodic orbits associated with the bending motion of the molecule, which result in characteristic patterns in the spectrum and the quantum number dependence of the vibrational fine structure spectroscopic constants. Two polar opposites are employed to elucidate the spectral patterns: the exact solution of the Schrödinger equation, using an accurate ab initio potential energy surface and an effective or resonance Hamiltonian (expressed in a harmonic oscillator basis set and block diagonalized into polyads), which is defined by parameters adjusted to fit either the measured or the calculated vibrational energies. The combination of both approaches together with nonlinear classical mechanics and semiclassical analyses provided a detailed spectroscopic picture of the breaking of one bond and the formation of a new one.

Further studies for HOCl, HOBr, HCN, CH$_2$ and ground and excited electronic states of O$_3$ showed that cascades of CS bifurcations of PO pave the road to dissociation or isomerization, as the molecule is excited along the reaction coordinate [6]. Two molecules are discussed here, hydrogen hypochlorite (HOCl) in its electronic ground state and the excited state of nitrous oxide (N$_2$O).

6.1.1 Hydrogen Hypochlorite

As an example of what we learn from a periodic orbit analysis of a highly excited triatomic molecule we review a study of HOCl [35]. A complete quantum mechanical calculation has been carried out for this molecule. Accurate high level quantum chemistry calculations have produced an analytical potential function valid for the global nuclear configuration space. Then, the nuclear Schrödinger equation is solved in Jacobi coordinates, the distance of Cl atom from the center of mass of OH, R, the bond length of OH, r, and the angle between the distances, γ, to produce hundreds of vibrational eigenstates with zero total angular momentum.

The eigenfunctions are visually examined to find out regularities and the degree of localization in the configuration space. As energy increases, the assignment becomes cumbersome, since, most of the wavefunctions show a complicated nodal structure. However, overtone states may appear regular at even very high energies, and thus,

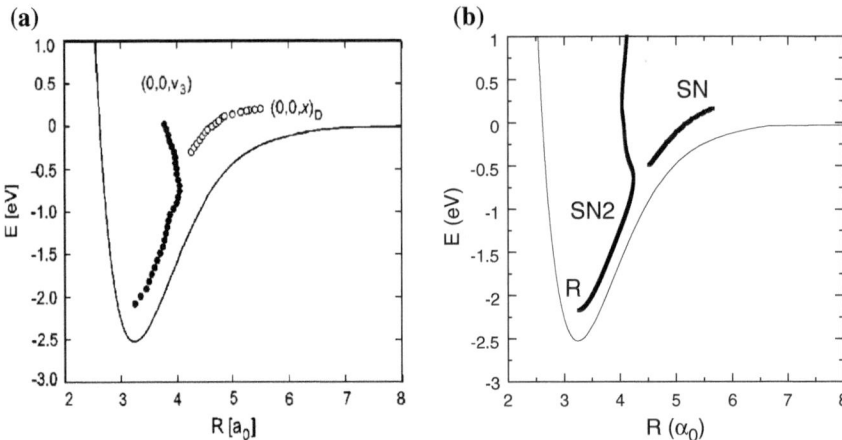

Fig. 6.1 a Minimum energy path for HOCl along the dissociation coordinate R; the potential is minimized in the other two degrees of freedom. The symbols indicate the energy and the extension of the wavefunctions in two overtone progressions of eigenstates $(0, 0, v_3)$ and $(0, 0, x)_D$, respectively. **b** Minimum energy path of HOCl along the dissociation coordinate R; the potential is minimized in the other two degrees of freedom. The *bold lines* indicate the maximum extension of the periodic orbits in R. R denotes the principal family of PO, and SN2 and SN center-saddle (saddle-node) bifurcations, respectively [35]

they become easily assignable. Most interesting is the normal mode overtones which lead the molecule to dissociation, and for HOCl this is the R mode.

It was found, that while initially the eigenfunctions are localized along the R coordinate, at some energy they started to deviate from this route. Simultaneously, a new progression of eigenfunctions emerged which were localized and properly oriented towards to the dissociation channel.

In Fig. 6.1a we show the overtone states of the R mode plotted on the minimum potential energy path. The circles depict the energy of the eigenstates, and also the extension of wavefunctions (localization). We can see, that at energy of about -0.5 eV the initial normal mode series, $(0, 0, v_3)$, diverges and a new series of eigenfunctions localized along the R coordinate appears, $(0, 0, x)_D$. v_3 and x are the number of quanta in the R mode. A similar analysis with periodic orbits is shown in Fig. 6.1b. The curves depict the energy of PO and the maximum extension in R. The principal family which corresponds to the R stretch deviates at about -0.5 eV and a new family appears after a center-saddle bifurcation. The continuation/bifurcation diagram of PO is depicted in Fig. 6.2a and the good correspondence among PO and overtone wavefunctions in Fig. 6.2b. For more details we address the reader to the review articles [13, 15, 16].

Another approach to study nonlinear phenomena in molecules is by employing spectroscopic Hamiltonians fitted to reproduce part of an experimental or theoretical spectrum. Then, by using Hamiltonian normal form expansion and semiclassical quantization the correspondence between classical and quantum mechanics is achieved [18, 28].

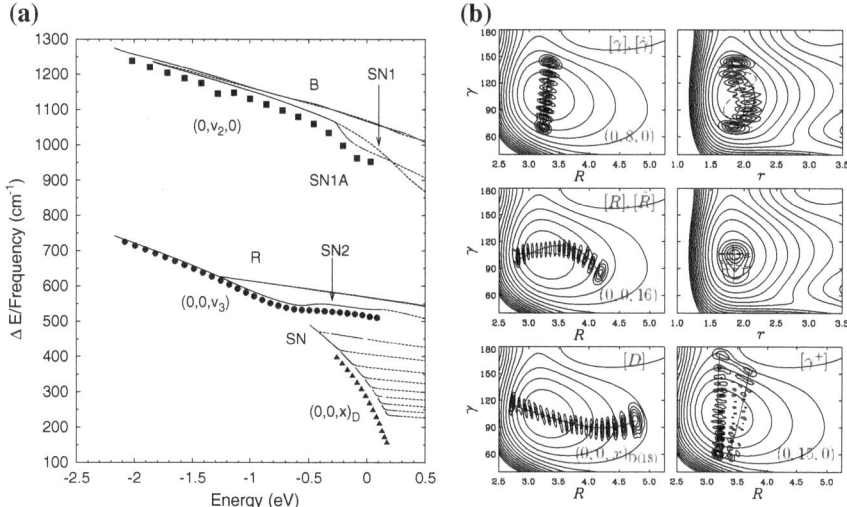

Fig. 6.2 **a** Continuation/Bifurcation diagram of PO (*continuous lines*) and quantum mechanical eigenlevels (*filled circles, squares and triangles*) of HOCl, **b** eigenfunctions and periodic orbits of HOCl [35]

6.1.2 Nitrous Oxide

The diffuse vibrational bands observed in the ultraviolet photodissociation spectrum of nitrous oxide (N_2O) by exciting the molecule in the first $^1A'$ electronic state, have recently been attributed to resonances localized mainly in the NN stretch and bend degrees of freedom. The origin of this localization was investigated by locating the fundamental families of periodic orbits emanating from several stationary points of the $^1A'$ potential energy surface as well as bifurcations of them. It was demonstrated that center-saddle bifurcations of periodic orbits are the main mechanism for creating stable regions in phase space that can support the partial trapping of the wavepacket, and thus, they explain the observed spectra. A nonlinear mechanical methodology, which involved the calculation of equilibria, periodic orbits and transition states in normal form coordinates, was applied for an in detail exploration of phase space. The fingerprints of the phase space structures in the quantum world were identified by solving the time dependent Schrödinger equation and calculating autocorrelation functions [19].

6.1.2.1 Equilibrium Points and Periodic Orbits

A portrait of the ab initio potential energy surface is shown in Fig. 6.3 in Jacobi coordinates (R the distance of O from the center of mass of NN, r the NN bond length and γ the angle between R and r). Eight stationary points were found and

Fig. 6.3 Excited potential
energy surface of N_2O (A^1A')
($\gamma = 0$) in Jacobi
coordinates. The *black bullet*
depicts the initial position of
the quantum wavepacket

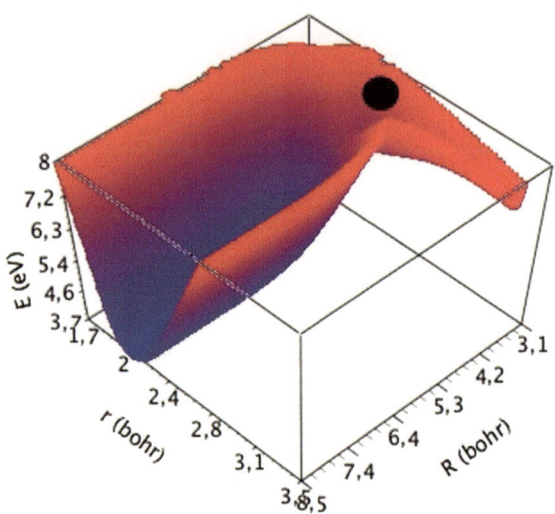

this demonstrates how complex a PES may become for an electronically excited
state, even for a triatomic molecule. This is the result of avoided crossings with
higher/lower electronic states.

Continuation/bifurcation (C/B) diagrams have been computed by locating families
of periodic orbits that emanate from the minimum EP1 (stable), and the saddles EP3
(index-1) and EP5 (index-2). These equilibria are close to the Franck-Condon region
in the photoexcitation of the molecule, and thus, they are related to the dissociation
dynamics of N_2O. Figure 6.4 shows a projection of the C/B diagram of periodic orbits
in the total energy—period plane, which emanate from the minimum EP1 (black
lines) and the saddle EP3 (red lines). Center—saddle (CS) bifurcations, which we
have associated with the diffuse bands of the photoabsorption spectrum of nitrous
oxide [26] are also shown with blue lines. m_i, $i = 1-3$, denote the fundamental
families emanated from the minimum, m_2^2 a period doubling bifurcation of m_2 and
CS_1 a CS bifurcation which is continued up to the spectroscopic excitation energies.
In Fig. 6.5 we show representative periodic orbits of the CS_i ($i = 2-4$) type families
projected on three different Jacobi coordinate planes together with equipotential
contours. $2S_R$ and $2S_r$ are the two fundamental families of EP3 equilibrium. The
plateau observed in the $2S_R$ family at about $54{,}280\,\mathrm{cm}^{-1}$ indicates a period doubling
bifurcation.

6.1.2.2 Normal Form Hamiltonians

To further investigate the dynamics of the molecule we have calculated the NHIM
around the EP3 equilibrium point (EP). Details of how to compute normal forms
(NF) are given in Sect. 5.5.3 and in the Appendix A.10.

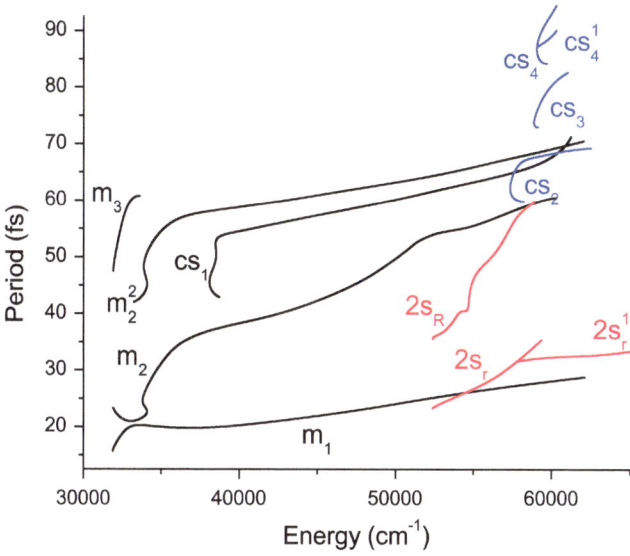

Fig. 6.4 A projection of the continuation/bifurcation diagram on the energy—period plane for families of periodic orbits, that emanate from the minimum of the $^1A'$—PES, EP1, (*black lines*), and the saddle EP3 (*red*). Center—saddle (CS) bifurcations, which are associated with the diffuse structure of the photoabsorption spectrum of nitrous oxide are also shown with *blue lines*. m_i, $i = 1-3$ denote the fundamental families emanated from the minimum, m_2^2 a period doubling bifurcation of m_2 and CS$_1$ a center—saddle bifurcation associated with the minimum. $2S_R$ and $2S_r$ are the fundamental families of EP3 equilibrium [19]

The expansion of the global Hamiltonian in Jacobi coordinates as a Taylor series is obtained by a least square fitting of the ab initio PES in a dense grid of points around the equilibrium and using spline interpolation to produce intermediate values. Another method could be the interpolation of the potential function at the equilibrium with Lagrange polynomials [8]. Then, we can evaluate all derivatives at any order by using Fornberg's algorithm. The order of the truncation and the accuracy of the NF expansion depend on the type of applications we want to do. Several accuracy criteria have been proposed [31, 32] and all of them are based on our ability to inversely transform from normal form to the natural (internal) coordinates. For the present case which includes an ab initio PES fitted by polynomials, we compare potential energy curves along the Jacobi coordinates by selecting points from the NF Hamiltonian. The normal form Hamiltonian is used to find initial conditions for the periodic orbits and this also provides another test for the accuracy of the NF expansion.

The order of the normal form Hamiltonian must be at least equal to the order of the Taylor expansion of the original Hamiltonian, otherwise the accuracy of the Taylor expansion is lost by not taking into account high order terms in the normalisation process. We have examined expansions up to tenth order. However, a Taylor expansion up to sixth order was employed for the EP3 equilibrium point and up to fourth order for EP5. Of course, convergence is not uniform for all degrees of freedom

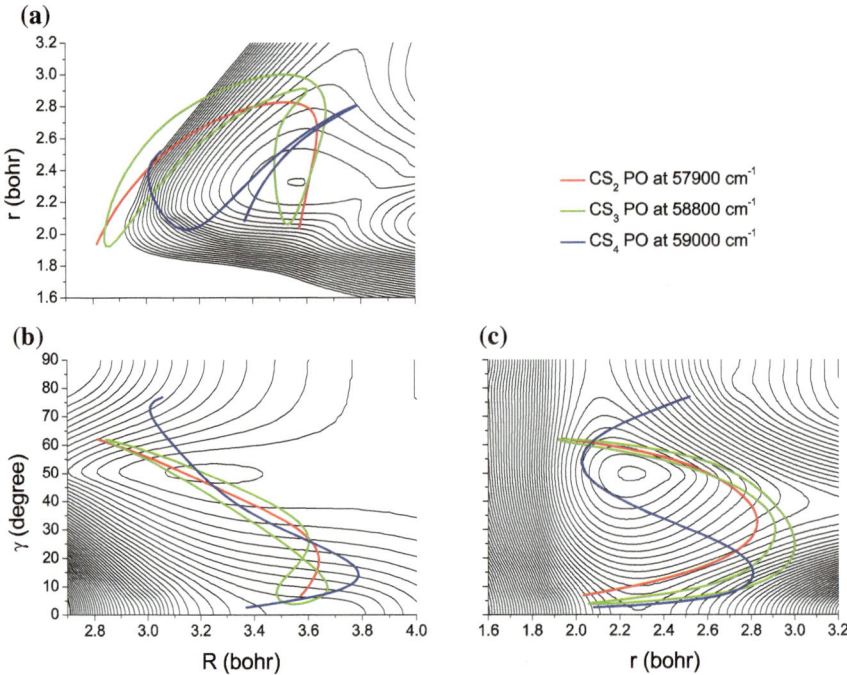

Fig. 6.5 Representative periodic orbits of CS_2 (*red*), CS_3 (*green*) and CS_4 (*blue*) families projected on the three Jacobi coordinate planes together with equipotential contours [19]

with that along R to be the worse. Comparison with the global potential gave good agreement up to energies about $1{,}000\,\text{cm}^{-1}$ above the equilibrium. Including terms higher than six and four respectively, the polynomial was improved at larger energies but deteriorated at lower. Since, we want high accuracy at energies close to equilibria we decided to truncate the expansion up to the orders mentioned above.

For a three degrees of freedom system the NHIM of EP3 equilibrium is three dimensional, whereas its stable and unstable manifolds ($W^{s(u)}$) four dimensional. The backward transformation described in Eq. A.119 helps to transform selected points in normal form coordinates back to Jacobi coordinates, whereas the forward transformation, Eq. A.118, maps phase space points in Jacobi representation to normal form coordinates.

Trajectories initialized properly on the (un)stable manifolds of the NHIM (W^u, W^s) are integrated forward and backward in time to produce the unstable and stable manifolds, Fig. 6.6. Description and details about forward-backward transformations can be found in references [29, 31, 36, 37]. As these authors have demonstrated the calculation of the stable and unstable manifolds of the NHIM allows the exact selection of reactive and non-reactive trajectories in normal form coordinates, which can then be transformed to the internal coordinate system. Figure 6.7 depicts the manifolds of NHIM projected on the nuclear configuration space. Because of the symmetry of the potential at $\gamma = 0$ the unstable degree of freedom is along γ.

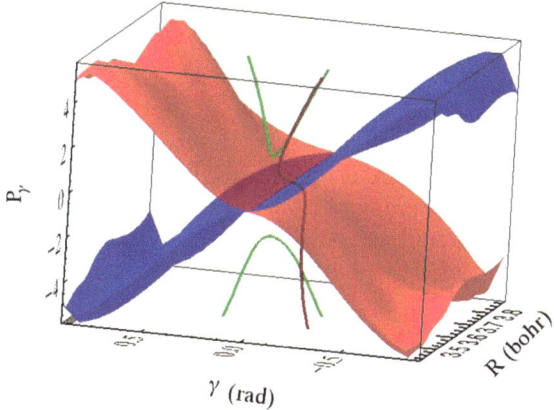

Fig. 6.6 The stable and unstable manifolds of the NHIM projected in the (R, γ, P_γ) Jacobi space. Three representative trajectories, one reactive (*red*) and two non-reactive (*green*) are also shown [19]

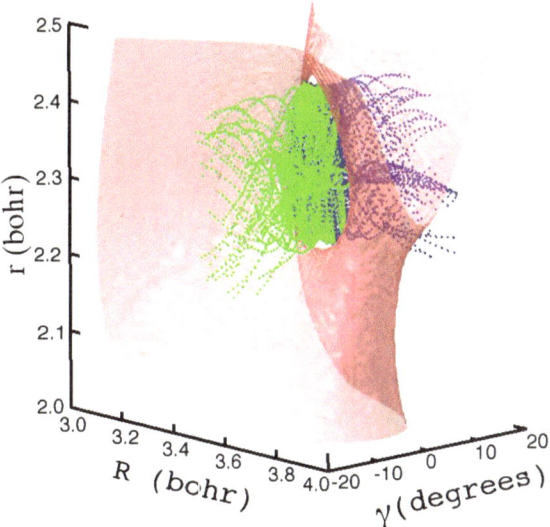

Fig. 6.7 The stable (*green*) and unstable (*blue*) manifolds of the NHIM shown in Fig. 6.6 projected in the Jacobi coordinate space [19]

6.1.2.3 Globalization via Periodic Orbits

Normal form Hamiltonians provide a local representation of the system and for energies not far away from the equilibrium point. Particularly, for complicated topologies such as the excited nitrous oxide, the PES cannot be described satisfactorily by a Taylor series expansion far from the equilibrium. Thus, Waalkens et al. [31] have proposed a globalization procedure to explore phase space regions far from the neighbourhood of the EP by propagating trajectories with initial conditions selected from the NF Hamiltonian, and then, transforming them back to the physical coordinate system. Still, with this method we are restricted to energies where the NF Hamiltonian is valid, i.e., close to equilibrium points. On the other hand, PO families generated by continuation methods explore the phase space at higher energies. Normal form Hamiltonians close to an EP are still useful, since, they can provide initial conditions for the first PO needed to start the continuation of the family. This method is important for high index saddles, where the first periodic orbit needed for the continuation of the family is difficult to obtain, in contrast to the PO of a NF Hamiltonian which are easily located.

A normal form Hamiltonian can also be formulated around a periodic orbit considering it as an equilibrium point of a Poincaré map, i.e., by defining a surface of section. However, the procedure to expand the Poincaré map in a Taylor series around the equilibrium is more cumbersome, since, it requires the simultaneous integration of a large number of trajectories in order to evaluate the high order derivatives. Similarly to equilibria, a normal form expansion will allow us to approximate reduced dimension tori as well as (un)stable manifolds around PO beyond the quadratic terms.

6.1.2.4 Quantum Mechanical Calculations

Wavepackets launched on periodic orbits at selected energies are propagated in time by solving the Schrödinger equation

$$\hat{H}\Psi(q, t) = \imath\hbar\frac{\partial\Psi(q, t)}{\partial t},\tag{6.1}$$

where $\imath = \sqrt{-1}$, $\hat{H}$ the Hamiltonian operator and $\Psi(q, t)$ is the wavefunction at time t. In Jacobi coordinates and for a rotationless molecule the Hamiltonian takes the form

$$
\begin{aligned}
\hat{H} = &-\frac{\hbar^2}{2\mu_R R}\frac{\partial^2}{\partial R^2}R - \frac{\hbar^2}{2\mu_r r}\frac{\partial^2}{\partial r^2}r \\
&-\frac{\hbar^2}{2}\left(\frac{1}{\mu_R R^2} + \frac{1}{\mu_r r^2}\right)\frac{1}{\sin\gamma}\frac{\partial}{\partial\gamma}\sin\gamma\frac{\partial}{\partial\gamma} + V(R, r, \gamma),
\end{aligned}\tag{6.2}
$$

where μ_R and μ_r are the reduced masses for O-N_2 and NN, respectively. Recurrences of the wavepacket to the initial region are traced by calculating the autocorrelation function, $C(t) = \langle\Psi(t = 0)|\Psi(t)\rangle$ (see Sect. 4.2). The Fourier transform of the autocorrelation function yields the absorption spectrum [24] and it has been discussed before [25, 26].

To propagate wavepackets in time we used the parallel code GridTDSE [27] (see also Sect. 5.3.1), which can be applied to general Cartesian coordinate systems. The

separation and conservation of the total angular momentum is obtained not by transforming the Hamiltonian in internal curvilinear coordinates but instead, by keeping
the Cartesian formulation of the Hamiltonian operator and projecting the initial wavefunction onto the proper irreducible representation angular momentum subspace. The
increased number of degrees of freedom by three, compared to previous methods for
solving the Schrödinger equation, is compensated by the simplicity of the kinetic
energy operator and its finite difference expressions [9, 10] which results in sparse
Hamiltonian matrices.

We have computed autocorrelation functions with initial wavepackets launched
on periodic orbits. The calculations have been carried out with a grid $(R, r, \gamma) \rightarrow$
$(161, 230, 100)$ points and $\gamma \in [0, 90]$ degrees and propagating the wavepacket for
$130\,fs$ [19].

6.2 Biological Molecules and Empirical Potentials

Biomolecules are complex systems, and therefore, it is not surprising that statistical
mechanical methods are employed for their study. The systematic methods of nonlinear mechanics based on hierarchically calculating stationary objects such as periodic
orbits, tori and stable and unstable manifolds are considered only for systems with
a few degrees of freedom. However, we argued before that periodic orbits offer the
means to extract the physics from complicated calculations, and even to get reliable
estimates of eigenenergies. Indeed, we have demonstrated that periodic orbits can
be located in biomolecules such as the dipeptide of alanine, a molecule with sixty
internal degrees of freedom [4].

6.2.1 Alanine Dipeptide

Empirical potential functions for biomolecules are usually constructed with pair addi-
tive potentials. Chemical bonds are described with harmonic as well as anharmonic
potentials, such as Morse type functions. For angles periodic functions are employed,
whereas intermolecular interactions are described by Lennard-Jones and Coulomb
potentials suitable for interactions among charged species. Potential functions as
those appeared in Eqs. 6.3–6.7 are the most commonly found in the literature [23].
The parameters in these functions are fitted to experimental as well as to theoretical
results.

$$V(R) = V(R)_{internal} + V(R)_{external} \tag{6.3}$$

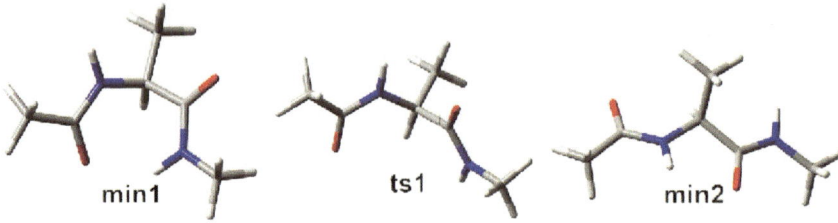

Fig. 6.8 The geometries of the two lowest minima of alanine dipeptide (2-acetamido-N-methylpropanamide, $CH_3CONHCH(CH_3)CONHCH_3$) at -16.53 kcal/mol (*min*1), -15.59 kcal/mol (*min*2), and the index-1 saddle at -15.00 kcal/mol (*ts*1). Oxygen atom is displayed in *red*, nitrogen in *blue*, carbon in *gray* and hydrogen in *white* [4]

$$V(R)_{internal} = \sum_{bonds} D_b[\exp(-2\alpha_b x) - 2\exp(-\alpha_b x)]$$

$$+ \sum_{angles} K_\theta(\theta - \theta_0)^2 + h.o.t.$$

$$+ \sum_{dihedrals} K_\chi[1 + \cos(n\chi - \sigma)] \tag{6.4}$$

$$V(R)_{external} = V(R)_{LJ} + V(R)_C \tag{6.5}$$

$$V(R)_{LJ} = \sum_{nonbonding} \varepsilon_{ij} \left[\left(\frac{R_{min,ij}}{r_{ij}} \right)^{12} - \left(\frac{R_{min,ij}}{r_{ij}} \right)^6 \right] \tag{6.6}$$

$$V(R)_C = \sum_{nonbonding} \frac{q_i q_j}{\varepsilon_D r_{ij}} + m.e. \tag{6.7}$$

The potential of alanine dipeptide has been constructed by using the parameters of the Charmm27 force field [7], Morse functions for the bond stretches and harmonic potentials for the angles. For a sixty degrees of freedom molecule like alanine dipeptide, the number of stationary points found is large. Here, we focus on the isomerization process between the two lowest minima. Figure 6.8 depicts the geometries of the two minima (*min*1 and *min*2) and the configuration of the index-1 saddle (*ts*1). We have located the fundamental families of periodic orbits,[1] $f23$ and $f24$, which are related to the isomerization of dipeptide [4]. It was found, that contrary to our expectation due to the small barrier, domains of phase space where trajectories are trapped for tens of picoseconds even at high excitation energies can be traced from these stationary points.

---

[1] At the vicinity of the equilibrium point the fundamental (or principal) families of periodic orbits are in one-to-one correspondence with the harmonic normal modes.

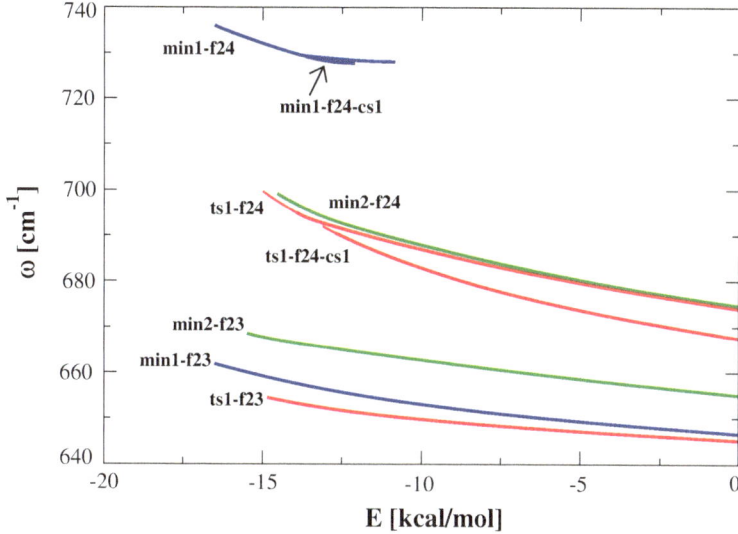

Fig. 6.9 Continuation/bifurcation diagram of alanine dipeptide in energy-frequency plane for the three equilibria studied [4]

In Fig. 6.9 the continuation/bifurcation diagram for the $f23$ and $f24$ fundamental families of periodic orbits originated from the three equilibria of the molecule is shown. The anharmonic behaviour of the vibrational modes is evident. For the $f24$ mode of $min1$ and $ts1$ equilibrium points a center-saddle bifurcation is observed. This means that at a specific energy the continuation line levels off, decreasing its anharmonicity, and a new pair of families of periodic orbits emerge, one of them with stable periodic orbits and the other with unstable ones (we show the stable branch).

We have visually examined the motions of the atoms for all families of PO at several energies. We confirmed that the $f23$ and $f24$ modes are mainly local type motions even at high excitation energies. By minimizing the energy starting from phase space points along the periodic orbits, we found that points in the region of $f23$ mode lead to $min1$, whereas by quenching from the region of $f24$ mode the system converges to $min2$.

The present study unequivocally demonstrates the existence of stable periodic orbits for substantial energy ranges in alanine dipeptide described with an empirical potential function. Empirical potentials are not unique, which means, that other functions can reproduce the data used to adjust the parameters. However, the advantage of locating stationary classical objects, such as periodic orbits, is proved by the expected *structural stability* of these objects. This means, that small perturbations either in the potential function of the molecule or its environment will not introduce major topological changes but only small differences in the numbers. In other words, center-saddle bifurcations will continue to exist in the perturbed system. Although, one has also to prove that localization remains in quantum calculations, the previous work on small molecules [13] supports our expectations that such phenomena will remain in the quantum world.

We have found [4] different times in the isomerization process depending on the excitation of specific vibrational mode and for different conformations. In spite of exciting similar modes in the three conformations their dynamics differ substantially. Controlling chemical reactions at such a level is one of the goals of chemical dynamics. For example, novel spectroscopic methods have indeed appeared which study small peptides in subpicosecond time scale. In an investigation of alanine tripeptide in water by two-dimensional vibrational spectroscopy, conformational fluctuations at the time scale of 0.1 ps have been reported [38].

6.2.2 Active Site of Cytochrome C Oxidase

A model for the active site of oxoferryl intermediate of cytochrome c oxidase has been constructed with 95 atoms [2]. Density functional theory has been employed to find equilibrium structures, harmonic frequencies and charge distributions for the active site. These data together with the force fields of Amber99 [33] and Charmm27 [7] have been used to construct an empirical potential function for the active site. The resulting analytical potential employs Morse type functions for the stretches, harmonic potentials for the angles, cosine functions for the torsions, and Lennard-Jones and Coulomb intermolecular interactions [23]. The geometry at equilibrium is shown in Fig. 6.10.

In Fig. 6.11 we plot the C/B diagram of periodic orbits in energy-frequency projection plane for those fundamental families that are mainly associated with the heme $a_3-Fe^{IV}=O$ species. The labels correspond to the harmonic normal modes numbering from which the family originates [20, 34]. We can see three major frequency regions. The low frequency f134 family corresponds to a breathing mode of the Imidazole in the proximal area of iron (Fig. 6.10). The f139 is associated with an oscillation of $Fe-N$ bond in the Imidazole$-Fe=O$ species, but it appears to be highly anharmonic. This anharmonicity results in a center-saddle bifurcation, cs139a.

The f139 family of PO is an example of how anharmonicity and coupling to other degrees of freedom can drastically change the harmonic vibrational frequency of the mode even with a small increase in energy. Unfortunately, it is difficult to foresee such nonlinear phenomena. Usually, spectroscopic investigations or detailed calculations are required.

The middle frequency families, f141, f142 and f143 differ only by a few wavenumbers and show small anharmonicities. They involve oscillations of Fe with the porphyrin ring. The higher frequency periodic orbits of the f145 family show slight negative anharmonicity and represent asymmetric oscillations of the Imidazole$-Fe=O$ species. In all these families the hydroxyl group attached to Cu_B shows appreciable displacements.

Examining the stability of periodic orbits we found that most of them remain stable for all energies studied, except the f139 family which turns to single unstable, i.e., only

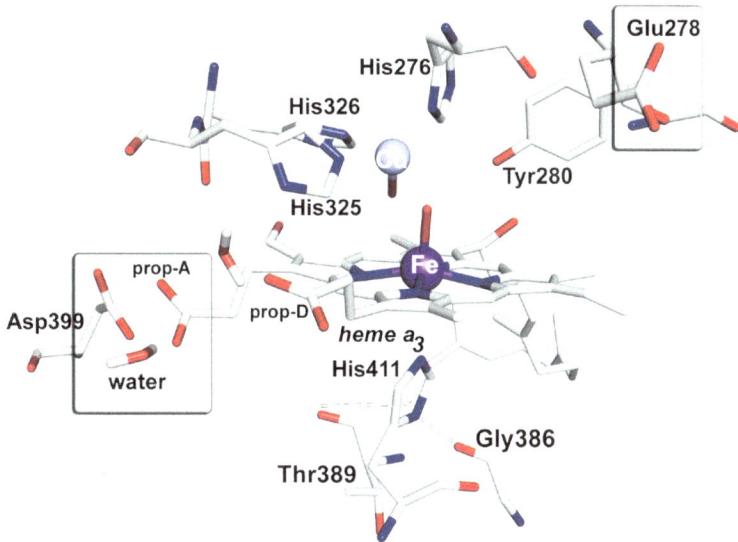

Fig. 6.10 Equilibrium conformation for the active site of the oxoferryl intermediate of cytochrome *c* oxidase. The model contains a Cu_B^{II} metal coordinated to two Imidazoles, a cross-linked Imidazole-phenol unit and a hydroxyl ($-OH$) group, and a heme a_3-Fe^{IV} *center*, in which the axial coordination of ferryl iron contains an Imidazole (replacing Histidine 411 according to the aa_3 Paracoccus denitrificans numbering) and an oxo (O^{2-}) ligand. The heme a_3 is taken without the propionate groups (A and D) and it is represented by an iron-porphyrin with only a CH(OH)CH3 group substituting the hydroxyethylfarnesyl side chain. This latter group interacts via a hydrogen bond with the cross-linked protonated phenol. Oxygen atom is displayed in *red*, nitrogen in *blue*, carbon in *gray* and hydrogen in *white* [2]

one degree of freedom is unstable. This means, that the nearby trajectories can escape through the unstable degree of freedom, and thus, explore larger regions of phase space. Stable periodic orbits trap the trajectories in their vicinity for considerably longer times, and thus, they contribute to coherent motions in the molecule.

To see how these periodic orbits affect the spectra we integrate trajectories by randomly selecting the velocities of atoms and scaling them accordingly to achieve an average temperature of 300 K. From a time series of 2.5 ns for Fe=O bond, we produce the power spectra shown in Fig. 6.12. For comparison we depict several curves. The absolute Fe=O spectra of the active site without constraints (c) (blue curve), keeping the Fe$-$N(Imidazole) distance constant at its optimized value (b) (red), and the Fe$-$N absolute spectrum of the active site when Fe=O distance is kept constant at its optimized value (a) (dark-yellow).

In Fig. 6.12 we can see that, by constraining the Fe$-$N (Imidazole) bond to its crystallographic value the 773 cm^{-1} band disappears (b, red curve), while constraining the Fe=O distance the band with a peak at 776 cm^{-1} is the dominant one in the frequency of Fe$-$N spectrum for the same spectral region (700–860 cm^{-1}) (a, dark-yellow). In accordance to the periodic orbit analysis, we assign the 773 peak to a

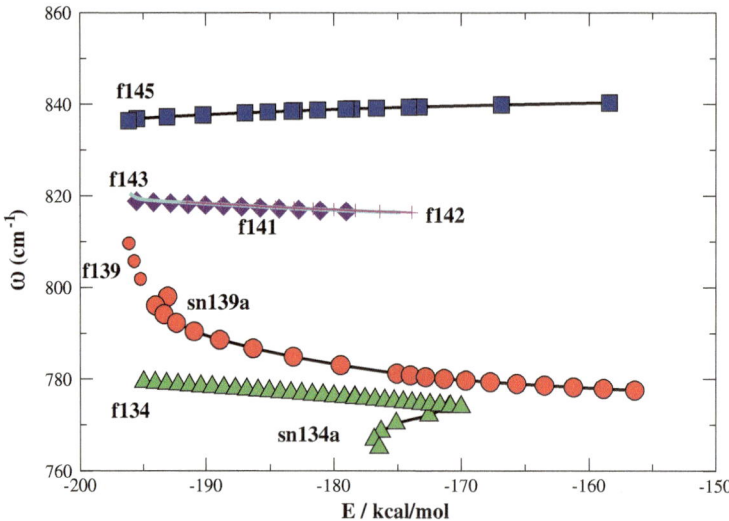

Fig. 6.11 Continuation/bifurcation diagram in the energy-frequency plane of the active site of cytochrome c oxidase [2]

breathing motion of Imidazole at the proximal region of iron. Comparing spectra of the total protein [2] with those of the active site we find a good agreement in the positions of the peaks at $798\,\mathrm{cm}^{-1}$, which demonstrates that the active site alone can reproduce vibrational frequencies of the enzyme. The smoothed absolute spectrum of the active site (c) exhibits three bands, 773, 810–820 and $840\,\mathrm{cm}^{-1}$, in accordance to the PO analysis. How these bands are affected by the protein environment and how they change in the protonated/deprotonated counterparts have been discussed in Ref. [2].

The sensitivity of the spectra on different protonated intermediate of oxofer-ryl found before led us to examine the spectra of the model system by varying the dielectric constant in the Coulomb potential. The dielectric constant varies in the range of [1–80] and the frequency moves from 823 to $796\,\mathrm{cm}^{-1}$[17]. This is the observed frequency range in resonance Raman spectra [30].

In the reaction of the mammalian mixed valence form of CcO with O_2, in which only heme a_3 and Cu$_B$ are reduced, the frequency at $804\,\mathrm{cm}^{-1}$ probed by resonance Raman spectroscopy has been attributed to Fe^{IV}=O bond. Contrary to that, in the reaction of the fully reduced enzyme with O_2 either two bands are identified with prominent frequencies at 790 and $804\,\mathrm{cm}^{-1}$ or only one [30] at $790\,\mathrm{cm}^{-1}$. Simulations and experimental spectra support the view that the one or two bands spectra is the result of the different Coulomb environment which may result from different protonation states of the enzyme.

Further support to these findings has come from quantum mechanics/molecular mechanics and molecular dynamics studies [3]. These highly elaborate calculations, supported by several calculations on smaller model systems, demonstrate the sensitiv-

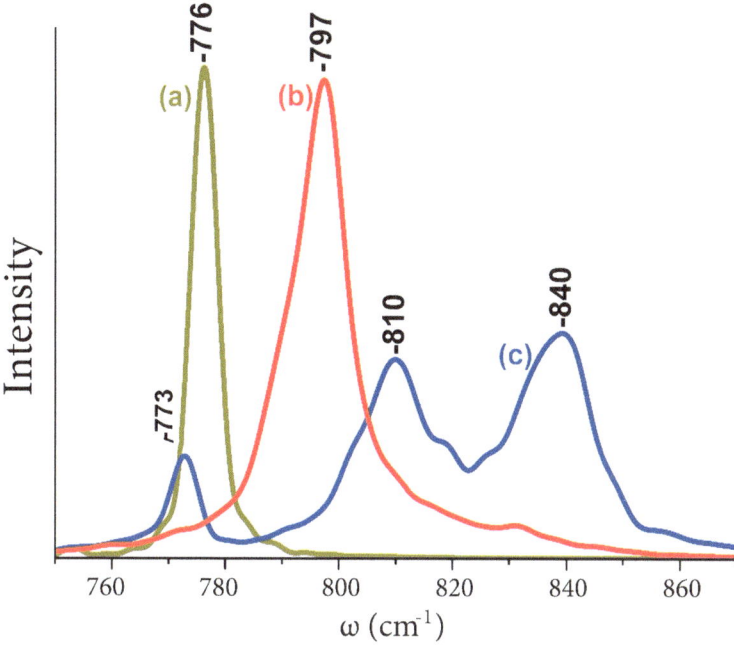

Fig. 6.12 Power spectra of the active site of cytochrome c oxidase: **a** (Fe−N(Imidazole)) absolute spectrum of the active site when Fe=O distance is kept constant at its optimized value (*dark-yellow*), **b** (Fe=O) absolute spectrum with Fe−Imidazole distance constant at its optimized value (*red*), and **c** spectrum without constraints (*blue curve*) [2]

ity of vibrational frequencies on the Coulomb field of heme a_3 and their dependence on the distance of the adjacent Cu_B to the heme a_3−Fe atom. This distance effect seems to be associated with the protonation state of the heme a_3 propionate A, and we proposed that it plays a crucial role on the function of CcO. More specifically, it was found a multiple (1:1:2) resonance among the frequencies of Fe^{IV}=O bond stretching, the breathing mode of Histidine 411, and a bending mode of the His411−Fe^{IV}=O species (aa_3 from Paracoccus denitrificans numbering). Furthermore, calculations on model systems demonstrated that the position of Cu_B in relation to heme a_3 iron-oxo plays a crucial role in regulating that resonance.

Summarizing, continuation/bifurcation diagrams of periodic orbits portray the geometry of phase space, and thus, they uncover nonlinear dynamical effects in complex biomolecules. Bifurcations of periodic orbit families indicate important resonances among vibrational modes which may have significant consequences in the dynamics and spectroscopy of the molecule. We have demonstrated that among the plethora of vibrational modes of model systems, such as the alanine dipeptide and the active site of oxoferryl intermediate of cytochrome c oxidase, center-saddle elementary bifurcations emerge from anharmonic fundamental modes which affect the isomerization process and the vibrational frequencies of the model molecules, respectively. Furthermore, for oxoferryl it is shown that this anharmonic frequency is also sensitive to the electrostatic environment, causing a shift in the vibrational frequency of N−Fe=O species.

The development of novel spectroscopic methods for the study of isolated molecules as well as in solutions at sub-picosecond times will increase the need for systematic theoretical investigations such as the periodic orbit analysis offers.

References

1. Dai HL, Field R (1995) Molecular dynamics and spectroscopy by stimulated emission pumping., Advanced series in physical chemistryWorld Scientific Publishing Company, Singapore
2. Daskalakis V, Farantos SC, Varotsis C (2008) Assigning vibrational spectra of ferryl-oxo intermediates of cytochrome c oxidase by periodic orbits and molecular dynamics. J Am Chem Soc 130:12,385–12,393
3. Daskalakis V, Farantos SC, Guallar V, Varotsis C (2010) Vibrational resonances and Cu_B displacement controlled by proton motion in cytochrome c oxidase. J Phys Chem B 114:1136–1143
4. Farantos SC (2007) Periodic orbits in biological molecules: phase space structures and selectivity in alanine dipeptide. J Chem Phys 126:175,101–175,107
5. Farantos SC, Qu Z, Zhu H, Schinke R (2006) Reactions paths and elementary bifurcations tracks: the diabatic $^1B_2$-state of ozone. Int J Bif Chaos 16:1913–1928
6. Farantos SC, Schinke R, Guo H, Joyeux M (2009) Energy localization in molecules, bifurcation phenomena, and their spectroscopic signatures: the global view. Chem Rev 109(9):4248–4271
7. Foloppe N, MacKerell AD (2000) All-atom empirical force field for nucleic acids: I. parameter optimization based on small molecule and condensed phase macromolecular target data. J Comput Chem 21:86–104
8. Fornberg B (1998) A practical guide to pseudospectral methods. Cambridge University Press, Cambridge
9. Guantes R, Farantos SC (1999) High order finite difference algorithms for solving the Schrödinger equation in molecular dynamics. J Chem Phys 111:10,827–10,835
10. Guantes R, Farantos SC (2000) High order finite difference algorithms for solving the Schrödinger equation in molecular dynamics. II. periodic variables. J Chem Phys 113:10,429–10,437
11. Hamm P, Lim M, DeGrado WF, Hochstrasser RM (1999) The two-dimensional IR nonlinear spectroscopy of a cyclic penta-peptide in relation to its three-dimensional structure. Proc Natl Acad Sci USA 96:2036
12. Herman M, Lievin J, Auwera JV, Campargue A (1999) Global and accurate vibration Hamiltonians from high-resolution molecular spectroscopy. Adv Chem Phys 108:1–431
13. Ishikawa H, Field RW, Farantos SC, Joyeux M, Koput J, Beck C, Schinke R (1999) HCP - CPH isomerization: caught in the act. Ann Rev Phys Chem 50:443–484
14. van der Meer J-C (1985) The Hamiltonian Hopf bifurcation. Springer, New York
15. Joyeux M, Farantos SC, Schinke R (2002) Highly excited motion in molecules: saddle-node bifurcations and their fingerprints in vibrational spectra. J Phys Chem 106:5407–5421
16. Joyeux M, Grebenshchikov SY, Bredenbeck J, Schinke R, Farantos SC (2005) Intramolecular dynamics along isomerization and dissociation pathways, in geometrical structures of phase space in multi-dimensional chaos. Adv Chem Phys 130:267–303
17. Kampanarakis A, Farantos SC, Daskalakis V, Varotsis C (2012) Non-linear vibrational modes in biomolecules: a periodic orbits description. Chem Phys 399:258–263
18. Main J, Jung C, Taylor HS (1997) Extracting the dynamics in classically chaotic quantum systems: spectral analysis of the HO_2 molecule. J Chem Phys 107:6577–6583

19. Mauguiére FAL, Farantos SC, Suarez J, Schinke R (2011) Non-linear dynamics of the photodissociation of nitrous oxide: equilibrium points, periodic orbits, and transition states. J Chem Phys 134(24):244,302–244,312

20. Moser J (1976) Periodic orbits near an equilibrium and a theorem by Alan Weinstein. Commun Pure Appl Math 29:727–747

21. Prosmiti R, Farantos SC (1995) Periodic orbits, bifurcation diagrams and the spectroscopy of C_2H_2 system. J Chem Phys 103(9):3299–3314

22. Prosmiti R, Farantos SC (2003) Periodic orbits and bifurcation diagrams of acetylene/vinylidene revisited. J Chem Phys 118(18):8275–8280

23. Rapaport DC (1995) The art of molecular dynamics simulation. Cambridge University Press, Cambridge

24. Schinke R (1993) Photodissociation dynamics. Cambridge University Press, Cambridge

25. Schinke R (2011) Photodissociation of N_2O: potential energy surfaces and absorption spectrum. J Chem Phys 134(064):313

26. Schinke R, Suarez J, Farantos SC (2010) Photodissociation of N_2O-frustrated NN bond breaking causes diffuse vibrational structures. J Chem Phys 133(091):103

27. Suarez J, Farantos SC, Stamatiadis S, Lathouwers L (2009) A method for solving the molecular Schrödinger equation in Cartesian coordinates via angular momentum projection operators. Comp Phys Comm 180:2015–2033

28. Svitak J, Li Z, Rose J, Kellman ME (1995) Spectral patterns and dynamical bifurcation analysis of highly excited vibrational spectra. J Chem Phys 102:4340–4354

29. Uzer T, Jaffé C, Palacián J, Yanguas P, Wiggins S (2002) The geometry of reaction dynamics. Nonlinearity 15:957–992

30. Varotsis C, Babcock GT (1990) Appearance of the νFe^{IV}=O) vibration from a ferryl-oxo intermediate in the cytochrome oxidase/dioxygen reaction. Biochemistry 29(32):7357–7362

31. Waalkens H, Burbanks A, Wiggins S (2004) Phase space conduits for reaction in multidimensional systems: HCN isomerization in three dimensions. J Chem Phys 121(13):6207–6225

32. Waalkens H, Schubert R, Wiggins S (2008) Wigner's dynamical transition state theory in phase space: classical and quantum. Nonlinearity 21:R1–R118

33. Wang J, Cieplak P, Kollman PA (2000) How well does a restrained electrostatic potential (resp) model perform in calculating conformational energies of organic and biological molecules? J Comput Chem 21:1049–1074

34. Weinstein A (1973) Normal modes for nonlinear Hamiltonian systems. Inv Math 20:47–57

35. Weish J, Hauschildt J, Düren R, Schinke R, Koput J, Stamatiadis S, Farantos SC (2000) Saddle-node bifurcations in the spectrum of HOCl. J Chem Phys 112(1):77–93

36. Wiggins S (1994) Normally hyperbolic invariant manifolds in dynamical systems. Springer, New York

37. Wiggins S, Wiesenfeld L, Jaffé C, Uzer T (2001) Impenetrable barriers in phase-space. Phys Rev Letters 86:5478–5481

38. Woutersen S, Mu Y, Stock G, Hamm P (2001) Subpicosecond conformational dynamics of small peptides probed by two-dimensional vibrational spectroscopy. Proc Natl Acad Sci USA 98:11,254–11,258

39. Xie A, van der Meer L, Hoff W, Austin RH (2000) Long-lived amide I vibrational modes in myoglobin. Phys Rev Letters 84:5435–5438

40. Zewail AH (2000) Femtochemistry: atomic-scale dynamics of the chemical bond. J Phys Chem A 104(24):5660–5694

Chapter 7
Epilogue

In the previous chapters the effort was focused to unveil the role of classical nonlinear mechanics in studying molecular dynamics by unfolding the geometry of phase space. This is obtained in a systematic way by locating time invariant structures, such as equilibria, periodic orbits, tori, normally hyperbolic invariant manifolds, stable and unstable manifolds. The hierarchical methodology of nonlinear mechanics to explore the structure of phase space is delineated in Fig. 1.4.

For decades tremendous effort has been put by chemists to construct molecular potential energy surfaces. The next step of locating the families of periodic orbits associated to particular stationary points on the PES is rarely executed, and that, in spite of the numerous and computationally expensive molecular dynamics calculations which usually follow the construction of PES. However, one must admit that by moving from the coordinate configuration space to phase space not only a deeper understanding of molecular dynamics is obtained, but furthermore, accurate classical mechanical calculations can be done by exploiting the local nature of the motions. The degree of the resolution in discovering phase space structures depends on the particular problem, and definitely, it can go beyond the experimental and quantum mechanical resolution.

Molecules are quantum objects, i.e., they obey the quantum mechanical laws. Solving the Schrödinger equation or an equivalent in other formulations of quantum mechanics, is restricted to small molecules and for a limited range of energies. What we have learnt by studying the phase space structure of several molecules, localized motions due to the existence of approximate constants of motion is very common even at high excitation energies. The ubiquitous center-saddle bifurcations of periodic orbits is a mechanism for creating regular regions embedded in chaotic sea, and thus, intriguing molecular spectroscopy and dynamics are elucidated or predicted. More importantly, by knowing the phase space structure approximate Hamiltonians (for example in normal forms) or empirical effective Hamiltonians based on spectroscopic data can be constructed to solve the quantum equations of motion or for carrying out a semiclassical quantization. In other words, by utilizing stationary phase space objects to form a *Grid* on which quantum mechanics is built, it turns out to be a promising method to study polyatomic molecules and for long time dynamics.

© The Author(s) 2014

S.C. Farantos, *Nonlinear Hamiltonian Mechanics Applied to Molecular Dynamics*,
SpringerBriefs in Electrical and Magnetic Properties of Atoms, Molecules, and Clusters,
DOI 10.1007/978-3-319-09988-0_7

Advances in computational chemistry have always been associated with progress in computers. As computational power increases larger molecular systems and more accurate electronic structure and molecular dynamics calculations are pursued. Quantum chemistry electronic structure and molecular dynamics computer codes have been developed almost independently of each other. Nowadays, the growth of parallel distributed computing [2] and high performance computers have made feasible the *on-the-fly* molecular dynamics, in which electronic structure calculations and integration of the classical equations of motion are executed simultaneously. Among these methods the most popular are the Born-Oppenheimer molecular dynamics (BOMD) [3]. These new algorithms, which unify Newton's and Schrödinger's equations, allow for complex simulations without relying on any adjustable parameter. *Ab initio* molecular dynamics simulations facilitate calculations on much larger molecules and time scales than previously thought feasible. With the knowledge obtained from a phase space analysis which reveals localization of the motions, we believe that BOMD calculations will become more efficient and accurate for even larger molecules and longer integration times.

To exploit present day computer technology suitable software is needed. Numerical calculations for both electronic structure and molecular dynamics are facilitated by employing Cartesian coordinates and discretizing space and time. We have shown, that variable order finite difference methods offer an unifying method for solving ordinary and partial differential equations. Nevertheless, the comprehension of physics and chemistry from lengthy computations on polyatomic molecules is not straightforward. The trajectories lie on non-Euclidean manifolds with variable dimensionality and this accents the importance of nonlinear mechanics. The interplay between quantum-classical theory and configuration-phase space manifold is the meaning that Fig. 2.1 conveys.

The necessity of approaching the theory of molecules by nonlinear mechanics is further stressed by the enormous progress in experimental techniques in the last decades. Methods based in ion-imaging technologies [1] allow one to excite reactant molecules and to characterize product molecules at specific quantum states, thus, enabling better control of chemical transformations. Such experimental techniques are expected to increase in the future with significant consequences to innovation in nanoscience, bio- and materials-science.

References

1. Ashfold MNR, Nahler NH, Orr-Ewing AJ, Vieuxmaire OPJ, Toomes RL, Kitsopoulos TN, Garcia IA, Chestakov DA, Wu SM, Parker DH (2006) Imaging the dynamics of gas phase reactions. Phys Chem Chem Phys 8:26–53
2. Foster I, Kesselman C (1999) The grid, blueprint for a new computing infrastructure. Morgan Kaufmann, San Francisco
3. Kühne TD (2013) Ab-initio molecular dynamics. arxiv:1201.5945v2

Appendix A
Calculus on Differentiable Manifolds

A.1 Topological Space

A topological space is a set T of arbitrary elements, the points, and a class of subsets U, called the *open sets* of T, such that the class contains the empty set ($\emptyset$) and itself (T), and it is *closed* under the formation of finite intersections ($U \cap V \cap \cdots \cap W$) and arbitrary unions ($U \cup V \cup \dots$). *The class of open subsets U "defines" the topology of T* [5, 9]. The above definition introduces the concept of *continuity* in the topological space by considering that every point of the set T has a *neighborhood*.

If for every pair of distinct points of T one can define neighborhoods that do not overlap, then this is a *Hausdorff* topological space.

Furthermore, if we can find a collection B of countable open sets U_i, such that every open subset of the topological space is represented as the union of elements of B, then T has a *countable basis*.

The Euclidean space of dimension n, $\mathbb{R}^n$, is a topological Hausdorff space with a countable basis. For $\mathbb{R}^n$ every neighborhood U of a point p there is an open set $U_i \subset U$.

A topological space is *compact* if from every *covering* of T by open sets, we can find a *finite* number of sets that still cover T.

A topological space is *connected* if it *cannot* be represented as the union of two or more *disjoint non-empty* open subsets.

© The Author(s) 2014

S.C. Farantos, *Nonlinear Hamiltonian Mechanics Applied to Molecular Dynamics*,
SpringerBriefs in Electrical and Magnetic Properties of Atoms, Molecules, and Clusters,
DOI 10.1007/978-3-319-09988-0

A.2 Differentiable (Smooth) Manifold

A manifold M of dimension m is a topological Hausdorff space with a countable basis, for which we can define a *chart* (a local coordinate system), i.e., a homeomorphism[1] (ϕ) (Fig. A.1),

$$\phi : U \subset M \to \phi(U) \subset \mathbb{R}^m, \tag{A.1}$$

of an open set U of M onto an open set $\phi(U)$ of $\mathbb{R}^m$. Since, the map is on an Euclidean space, we can also define a coordinate representation in $\mathbb{R}^m$

$$q^i = f^i \circ \phi \quad \text{or} \quad \phi(p) = (q^1(p), q^2(p), \ldots, q^m(p))^T \in \mathbb{R}^m, \tag{A.2}$$

for every point $p \in U$, and f^i are *smooth functions*,[2] i.e, they map the point $\phi(p)$ in the Euclidean space to real numbers[3] [3, 8, 9].

By defining a chart at every point $p \in M$ we form an *atlas*, that covers all the manifold M. If any pair of two charts overlap smoothly (*diffeomorphically*, i.e., there are smooth invertible transformations of one coordinate system to another (Fig. A.2), then we have a *differentiable or smooth manifold*.

For an Euclidean topological space, $\mathbb{R}^n$, a manifold $M \subset \mathbb{R}^n$ with dimension m, $0 < m < n$, is defined by the set of points $q = (q^1, q^2, \ldots, q^n)^T$, that satisfy a

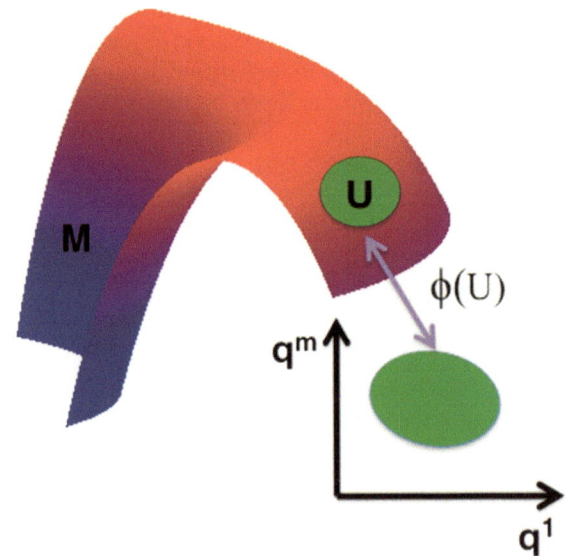

Fig. A.1 A coordinate chart is defined by the smooth (differentiable) function $\phi(U)$ of an open set $U \subset M$ at a point $p \in U$ of the manifold M onto an open set $\phi(U) \subset \mathbb{R}^m$

---

[1] A function ϕ between two manifolds is called a homeomorphism if it is bijective (one to one), continuous and with an inverse function, ϕ^{-1}, also continuous.

[2] A smooth function is loosely defined as a differentiable function as many times as it is required.

[3] The letter superscript (T) denotes a column vector and generally the transpose of a matrix.

Fig. A.2 A transition map which describes a diffeomorphic coordinate transformation, $F = \psi \circ \phi^{-1} : \phi(U \cap V) \mapsto \psi(U \cap V)$

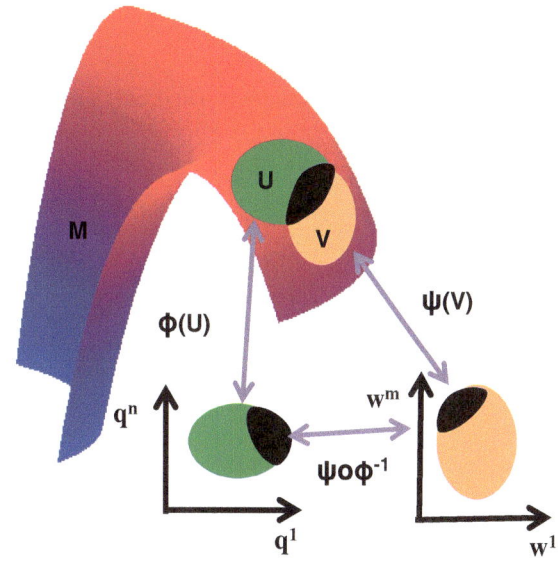

system of $k = n - m$ scalar equations: [4]

$$\phi^i(q^1, q^2, \ldots, q^n) = c^i, \quad i = 1, \ldots, k, \tag{A.3}$$

where ϕ are smooth functions, $\phi : \mathbb{R}^n \to \mathbb{R}^m$, $\phi = (\phi^1, \phi^2, \ldots, \phi^k)^T$, $q \in D \subset \mathbb{R}^n$ and $c = (c^1, c^2, \ldots, c^k)^T$, a column vector of constants.

The manifold M is *smooth (differentiable)*, if ϕ are smooth and the *rank* of Jacobian matrix $D_q \phi$ is equal to k at each point $q \in M$.

$$D_q\phi = \begin{bmatrix} \dfrac{\partial \phi^1}{\partial q^1} & \cdots & \dfrac{\partial \phi^1}{\partial q^{n-1}} & \dfrac{\partial \phi^1}{\partial q^n} \\ \dfrac{\partial \phi^2}{\partial q^1} & \cdots & \dfrac{\partial \phi^2}{\partial q^{n-1}} & \dfrac{\partial \phi^2}{\partial q^n} \\ \vdots & \vdots & \vdots & \vdots \\ \dfrac{\partial \phi^{k-1}}{\partial q^1} & \cdots & \dfrac{\partial \phi^{k-1}}{\partial q^{n-1}} & \dfrac{\partial \phi^{k-1}}{\partial q^n} \\ \dfrac{\partial \phi^k}{\partial q^1} & \cdots & \dfrac{\partial \phi^k}{\partial q^{n-1}} & \dfrac{\partial \phi^k}{\partial q^n} \end{bmatrix} \tag{A.4}$$

$k = n - m$ is the *codimension* of M, and equations $\phi^i = c^i$, $i = 1, \ldots, k$, give an *implicit representation* of the manifold M.

A.2.1 Implicit Function Theorem

Theorem A.1 (Implicit Function Theorem)

(i) **Implicit Representation - ϕ**

We consider that we are given: $\phi : D \to \mathbb{R}^m$, $D \subset \mathbb{R}^n$ $n > m$,

$$\phi = (\phi^1, \phi^2, \ldots, \phi^k)^T \in \mathbb{R}^k \qquad (A.5)$$

has

$$\phi(q^*) = c, \qquad (A.6)$$

for some $q = (q^1, q^2, \ldots, q^n)^T \in D$ and $c = (c^1, c^2, \ldots, c^k)^T \in \mathbb{R}^k$, $k = n - m$. If the Jacobian matrix $D_q\phi$ evaluated at a point q is non-singular, then near q, the level set of ϕ at c,

$$L_\phi(c) := \{\text{ all } q^* \in D, \text{ which satisfy } \phi(q^*) = c\}, \qquad (A.7)$$

is a $(m = n - k)$ - dimensional manifold embedded in D.

The tangent space (Sect. A.5) of this manifold at q^* is perpendicular to the row vectors of the matrix $D_q\phi$.

$$D_q\phi(q^*) = \begin{bmatrix} \partial_1\phi^1(q^*) & \cdots & \partial_{n-1}\phi^1(q^*) & \partial_n\phi^1(q^*) \\ \partial_1\phi^2(q^*) & \cdots & \partial_{n-1}\phi^2(q^*) & \partial_n\phi^2(q^*) \\ \vdots & \vdots & \vdots & \vdots \\ \partial_1\phi^{k-1}(q^*) & \cdots & \partial_{n-1}\phi^{k-1}(q^*) & \partial_n\phi^{k-1}(q^*) \\ \partial_1\phi^k(q^*) & \cdots & \partial_{n-1}\phi^k(q^*) & \partial_n\phi^k(q^*) \end{bmatrix} \qquad (A.8)$$

(ii) **Explicit Representation - χ**

The implicit function theorem can also be stated in the following way. There are smooth locally defined functions

$$q^{m+i} = \chi^i(q^1, q^2, \ldots, q^m), \quad i = 1, \ldots, k(= n - m), \qquad (A.9)$$

such that

$$\phi^i(q^1, q^2, \ldots, q^m, \chi^1(q^1, q^2, \ldots, q^m), \ldots, \chi^k(q^1, q^2, \ldots, q^m)) = c^i \qquad (A.10)$$

for all q in some neighborhood of $q^* \in D \subset \mathbb{R}^n$, which belong to the level set

$$L_\phi(c) := \{\text{ all } q^* \in D \text{ which satisfy } \phi(q^*) = c\}. \qquad (A.11)$$

We define the column vectors $q' = (q^1, \ldots, q^m)^T$, *and* $q'' = (q^{m+1}, \ldots, q^n)^T$.
The derivatives of χ *at* (q'^*) *are given by*

$$\frac{\partial \chi^i}{\partial q^j} = -\sum_{l=1}^{k} \left(\left[\frac{\partial \phi}{\partial \chi} \right]^{-1} \right)^i_l \left[\frac{\partial \phi^l}{\partial q^j} \right], \quad (i = 1, \ldots, k), \ (j = 1, \ldots, m).$$

$$(\text{A.12})$$

(iii) **Parametric Representation** - ψ

The parametric representation of the manifold M of dimension m assumes that the n coordinates of Euclidean space, $\mathbb{R}^n$, $n > m$, can be expressed as one-to-one functions $f = (f^1, f^2, \ldots, f^n), \in \mathbb{R}^n$ of the parameters $\psi = (\psi^1, \psi^2, \ldots, \psi^m)$, $\psi \in \mathbb{R}^m$

$$q = f(\psi^1, \ldots, \psi^m). \tag{A.13}$$

If we assume that the system

$$f(\psi) - q = 0, \tag{A.14}$$

can locally be solved with respect to $(\psi^1, \ldots, \psi^m, q^{m+1}, \ldots, q^n)$ as functions of the other variables, $(q^1, \ldots, q^m)$, then we have an explicit representation of the $m-$manifold in $(q^1, \ldots, q^m)$ coordinates

$$\psi^i = F^i(q^1, \ldots, q^m), \quad i = 1, \ldots, m$$
$$q^{m+i} = f^{m+i}[F^1(q^1, \ldots, q^m), \ldots, F^m(q^1, \ldots, q^m)]. \tag{A.15}$$

A.2.2 Inverse Function Theorem

Theorem A.2 (Inverse Function Theorem)
   If $f : D \to \mathbb{R}^n$, $D \subset \mathbb{R}^n$, $f = (f^1, f^2, \ldots, f^m), \in \mathbb{R}^m$ and $D_q f(q^*)$ is a linear isomorphism (Jacobian) for some $q^* \in D$, then

(i) *There exists a neighborhood U of q^* and a neighborhood V of $w^* = f(q^*)$ such that the restriction of f to U is an invertible function $f : U \to V$.*
(ii) *f^{-1} is differentiable at $w^* = f(q^*)$, and the derivative is given by*

$$D_w f^{-1}(w^*) = [D_q f(q^*)]^{-1}. \tag{A.16}$$

A.2.3　Example: 2D Torus in a 3D Euclidean Space

Fig. A.3 depicts a 2D torus.

- **Implicit representation**

$$\left(R - \sqrt{x^2 + y^2}\right)^2 + z^2 - r^2 = 0 \tag{A.17}$$

R is the distance from the　center of the tube to the center of the torus and r is the radius of the tube.

- **Explicit representation**

$$z = \pm \sqrt{r^2 - \left(R - \sqrt{x^2 + y^2}\right)^2} \tag{A.18}$$

- **Parametric representation**

$$x(u, v) = (R + r \cos v) \cos u \tag{A.19}$$
$$y(u, v) = (R + r \cos v) \sin u \tag{A.20}$$
$$z(u, v) = r \sin v \tag{A.21}$$

Fig. A.3 A 2D—torus embedded in a 3D Euclidean space

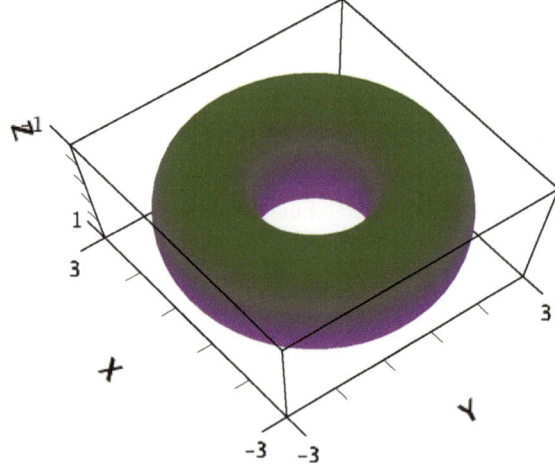

A.3 Smooth Functions and Maps on Manifolds

A smooth function f on a manifold M is a map from M to the real numbers $\mathbb{R}$,

$$f : M \to \mathbb{R} : p \in M \mapsto f(p) \in \mathbb{R}, \qquad (A.22)$$

which is differentiable. This means

$$f \circ \phi^{-1} : \phi(U) \subset \mathbb{R}^m \to \mathbb{R}, \qquad (A.23)$$

where (U, ϕ) is a chart with U an open set in the m-dimensional manifold M containing p. The set of all smooth functions on a manifold M is denoted by $\mathscr{F}(M)$.

We can also define a map F on a manifold M with dimension m to a manifold W of dimension n. If (U, ϕ) a chart on M and (V, ψ) a chart on W, such that if $p \in U \subset M$ then $F(p) \in V \subset W$, the coordinate transformation is the function $y = \psi \circ F \circ \phi^{-1} : \mathbb{R}^m \mapsto \mathbb{R}^n$ (Fig. A.4). If y is a smooth vector-valued function defined on an open set of $\mathbb{R}^m$ then F is a differentiable map at $p \in U \subset M$ with differential $DF = F_*$, a linear map (the Jacobian) of the tangent space $T_p M$ of M at the point p onto the tangent space $T_{F(p)} W$ at the point of $F(p)$ (Fig. A.10).

A *diffeomorphism* is a smooth mapping $F : M \mapsto W$ which has an inverse mapping, which is also smooth. The set of all diffeomorphisms $F : M \mapsto M$ is denoted as $Diff(M)$ and it is a group.

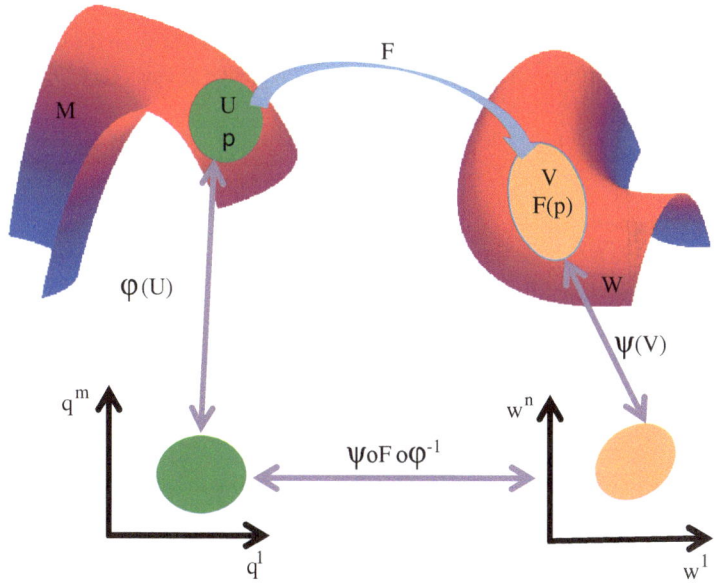

Fig. A.4 A differentiable map between manifolds

A.4 Curves on Manifolds

A smooth curve in an Euclidean space $\mathbb{R}^n$, $\gamma(t)$, is defined as a map, that leads from an open interval I of the real axis $\mathbb{R}$ to the $\mathbb{R}^n$,

$$\gamma : I \subset \mathbb{R} \rightarrow \mathbb{R}^n : t \in I \mapsto \gamma(t) \in \mathbb{R}^n. \tag{A.24}$$

By defining a coordinate system with a basis set e_i the curve and its derivative are described by the equations

$$\gamma(t) = \sum_{i=1}^{n} \gamma^i(t) e_i, \tag{A.25}$$

$$\frac{d\gamma(t)}{dt} = \sum_{i=1}^{n} \frac{d\gamma^i(t)}{dt} e_i, \tag{A.26}$$

or

$$\dot{\gamma}(t) = \sum_{i=1}^{n} \dot{\gamma}^i(t) e_i. \tag{A.27}$$

A.5 Tangent Vector Space

If the curve γ lies on the manifold M and at $t = 0$, $\gamma(0) = q_0$, then the derivative $\dot{\gamma}(0)$ is a *tangent vector* of M. Since, infinite curves may pass through the point q_0 we may say that their derivatives define a *tangent vector space* at q_0, $T_{q_0}M$ (Fig. A.5).

In the implicit representation of the manifold M with dimension m ($\phi^i = c^i$, $i = 1, \ldots, k = n - m$) embedded in an Euclidean space $\mathbb{R}^n$ of dimension n, a m -

Fig. A.5 If a coordinate chart is defined at a point $q_0 \in U$ of the manifold M, then the derivatives $\partial_i|_{q_0} \equiv \frac{\partial}{\partial q^i}|_{q_0}$ define a coordinate system in the tangent space $T_{q_0}M$

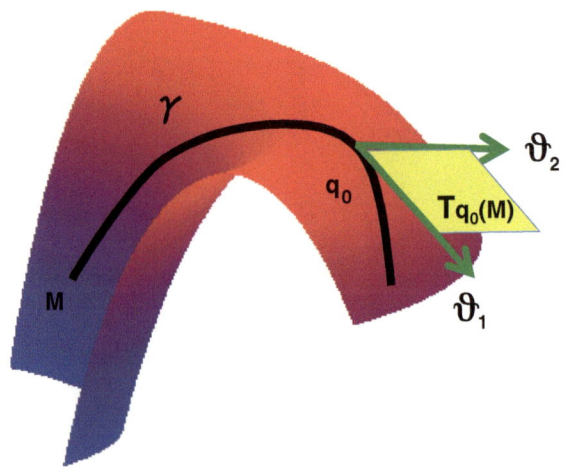

dimensional *tangent vector space*, $T_{q_0}M$ is defined and any vector v of the $T_{q_0}M$ is *perpendicular* to $\nabla\phi^i(q_0)$ (Theorem A.1).

The dimension of the tangent vector space is that of the manifold, m. Given a coordinate system (a chart) at the point q_0 with basis e_i, $i = 1, \ldots, m$, then the basis set for the tangent space is $\partial_i \equiv \frac{\partial}{\partial q^i}$, $i = 1, \ldots, m$,

$$v \in T_{q_0}M : \quad v = \sum_{i=1}^{m} v^i(q_0)\partial_i|_{q_0}. \tag{A.28}$$

If the (contravariant) vector $v \in T_{q_0}M$ acts on a smooth function $f \in \mathscr{F}(M)$, then

$$v(f) = \sum_{i=1}^{m} v^i \partial_i f(q_0). \tag{A.29}$$

Also,

Theorem A.3 *If $f \in \mathscr{F}(M)$ is a smooth function on the manifold M, then, there exists a unique vector, $\partial f(q_0) \in T_{q_0}M$, such that for all curves $\gamma \in M$ with $\gamma(0) = q_0$,*

$$\dot{f}(\gamma(t))|_{t=0} = \sum_{i=1}^{m} \dot{\gamma}^i(0)\partial_i f(q_0) \; = \; < \partial f(q_0)|\dot{\gamma}(0) >, \tag{A.30}$$

where $< | >$ denotes scalar (inner) product. $\partial f(q_0)$ is called intrinsic gradient of f at q_0.

If f is also defined on the n - dimensional Euclidean space $\mathbb{R}^n$, then, its gradient ($\partial f(q_0)$) has the *intrinsic derivative* ($\nabla f(q_0)$ to be distinguished from $\partial f(q_0)$) as its orthogonal projection on the tangent space at q_0 (Fig. A.6). That is, there are unique

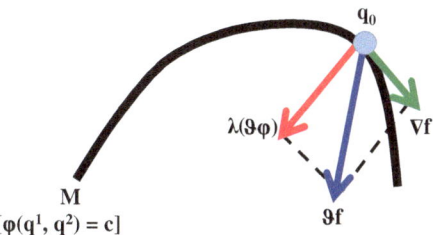

Fig. A.6 The manifold M has the implicit description of $\phi(q^1, q^2) = c$ with its gradient $\partial\phi$ at the point $q = q_0$ to be perpendicular to the intrinsic derivative, ∇f, of the function $f(q)$. The gradient of the function f at the point q_0, (∂f) is a 2D vector, but the intrinsic derivative lies in the 1D tangent space of the manifold. From the relation $(\partial f(q) - \lambda\partial\phi(q)) \bullet \partial\phi(q) = 0$ we take $\lambda = \frac{\partial f(q)\bullet\partial\phi(q)}{||\partial\phi(q)||^2}$. So, the intrinsic derivative is defined as $\nabla f(q) = \partial f(q) - \frac{\partial f(q)\bullet\partial\phi(q)}{||\partial\phi(q)||^2}\partial\phi(q)$

scalars $\lambda = (\lambda_1, \ldots, \lambda_k)$, such that

$$\nabla f(q_0, \lambda_1, \ldots, \lambda_k) = \partial f(q_0) - \sum_{i=1}^{k} \lambda_i \partial \phi^i(q_0). \tag{A.31}$$

λ_i, $i = 1, \ldots, k$, are called *Lagrange Multipliers*.

A.5.1 Extrema of a Function on a Manifold

The extrema of the function $f(q^1, \ldots, q^n)$ with n variables, which satisfy k constraints ($n > k > 0$)

$$\phi^i(q^1, \ldots, q^n) = 0, \ i = 1, \ldots, k, \tag{A.32}$$

are the extrema of f on the manifold which has the implicit representation given by Eq. A.32. Thus, we are seeking the extrema of the function

$$f^*(q, \lambda) = f(q^1, \ldots, q^n) - \sum_{i=1}^{k} \lambda_i \phi^i(q^1, \ldots, q^n), \tag{A.33}$$

i.e., the intrinsic derivative of f^* should be zero

$$\nabla f^*(q, \lambda) = \partial f(q) - \sum_{i=1}^{k} \lambda_i \partial \phi^i(q) = 0. \tag{A.34}$$

The *Lagrange Multipliers* λ_i are k unknown parameters to be determined by solving the system of equations Eqs. A.32 and A.34.

A.6 Tangent Bundle

At each point p of a m-dimensional manifold M we associate the tangent vector space $T_p M$. The set of all tangent spaces of the (differentiable) manifold form the *tangent bundle* of M, which is also a (differentiable) manifold,

$$TM = \bigcup_{p \in M} T_p M. \tag{A.35}$$

Fig. A.7 A schematic
(approximate) representation
of the tangent bundle of
manifold M

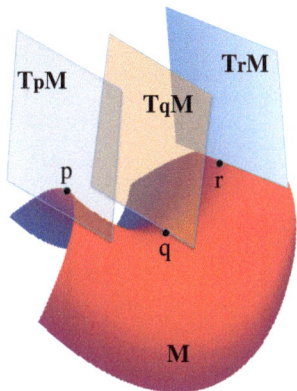

The bundle contains both the manifold M and its tangent spaces $T_p M$ called the *fibres* with M the *base space* (Fig. A.7). Thus, the dimension of the bundle is $dim(TM) = 2m$.

When we define charts (coordinates by the function ϕ) and an atlas on the manifold, then we can also define the tangent space of the chart

$$T\phi = TU \mapsto \phi(U) \times \mathbb{R}^m. \tag{A.36}$$

TM locally has the product structure $M \times \mathbb{R}^m$. However, its global structure may be more complicated.

A.7 Vector Field

A vector field V_F on a smooth manifold M is a map that assigns to every point p of M a *specific* tangent vector v_p taken from the vector space $T_p M$:

$$V_F : M \mapsto TM : p \in M \mapsto v_p \in T_p M. \tag{A.37}$$

The set of all vector fields of M is denoted as $\mathcal{V}_{\mathcal{F}}(M)$.

Vector fields acting on functions of $\mathcal{F}(M)$ are the *directional derivatives* of $\mathcal{F}$ (Sect. A.7.3).

A.7.1 Coordinate Transformation in Overlapping Charts

The base fields on two contiguous, overlapping open sets U and V with the charts $(\phi, U) \equiv (f^1 \circ \phi, \ldots, f^n \circ \phi)^T = (q^1(p), \ldots, q^n(p))^T$ and $(\psi, V) \equiv (w^1 \circ$

$\psi, \ldots, w^n \circ \psi)^T = (w^1(p), \ldots, w^n(p))^T$, respectively, are related as follows for a function $g \in \mathscr{F}(M)$ (Fig. A.2)

$$\frac{\partial(g \circ \phi^{-1})}{\partial f^i} = \sum_{k=1}^{n} \frac{\partial(g \circ \psi^{-1})}{\partial w^k} \frac{\partial(\psi^k \circ \phi^{-1})}{\partial f^i}, \tag{A.38}$$

or

$$\partial_i^\phi|_p(g) = \sum_{k=1}^{n} \partial_k^\psi|_p(g) \frac{\partial F^k}{\partial f^i}. \tag{A.39}$$

The matrix $\frac{\partial(\psi^k \circ \phi^{-1})}{\partial f^i}$ is the Jacobian matrix, $D_p F$, of the *transition map* $F = (\psi \circ \phi^{-1})$ (see Fig. A.2). In coordinates, we usually write $w^i = F^i(f^1, \ldots, f^n)$.

A.7.2 Local Flow

A local flow or a *local 1-parameter group of diffeomorphisms*, $\Phi_t(q)$, satisfies the differential equation

$$\frac{\partial}{\partial t} \Phi_t(q) \equiv \partial_t \Phi_t(q) \equiv \dot{\Phi}_t(q) = v(\Phi_t(q)), \tag{A.40}$$

where v is a vector field acting on the curve passing through q. Thus, Φ_t is an *integral curve* of the vector field v_q. It is proved that, for each point $q \in M$ there is precisely one integral curve Φ with initial point at $t = 0$, $\Phi_0 = q$.

For the times t, s and $t + s$ it is valid that

$$\Phi_t \circ \Phi_s = \Phi_{t+s} = \Phi_s \circ \Phi_t \tag{A.41}$$

A.7.3 Directional Derivative

The directional derivative of a function f on M is defined as the action of a vector field V_F on the function

$$V_F \in \mathscr{V}_{\mathscr{F}}(M) : f \in \mathscr{F}(M) \mapsto V_F(f) \in \mathscr{F}(M) \tag{A.42}$$

$$: f(p) \mapsto v_p(f) = \sum_i v^i \partial_i f. \tag{A.43}$$

In a chart (ϕ, U), a vector field can be represented locally by means of coordinate vector fields, or *base fields*. For every point p of an open neighborhood $U \subset M$, the base $\partial_i|_p$, is defined as a *vector field* on U:

$$\partial_i : U \mapsto TU : p \in U \mapsto \partial_i|_p. \tag{A.44}$$

So, if we denote the chart map by $\phi(p) = (q^1(p), \ldots, q^n(p))$, any smooth vector field V_F defined on $U \subset M$ has the local representation on U

$$V_F = \sum_{i=1}^{n} [V_F(q^i)]\partial_i = \sum_i v_F^i \partial_i, \tag{A.45}$$

where v_F^i the ith component of the vector field (see Fig. A.5).

A.7.4 Lie Derivative of Vector Fields

Let X and Y be pair of vector fields on a manifold M and let $\Phi_t = \Phi(t)$ be the local flow generated by the field X ($\dot{\Phi}_{tq} = X_q$). Then, Φ_{tq} is the point at time t along the *integral curve* of X, the *orbit* of q, that starts at time $t = 0$ at the point q. The Lie derivative of Y with respect to X is defined to be the vector field $\mathscr{L}_X Y$ whose value at q is

$$\begin{aligned}
[\mathscr{L}_X Y]_q &= \lim_{t \to 0} \frac{[Y_{\Phi_{tq}} - \Phi_{t*} Y_q]}{t}, \\
&= \lim_{t \to 0} \frac{[\Phi_{-t*} Y_{\Phi_{tq}} - Y_q]}{t}, \\
&= \left\{ \frac{d}{dt} (\Phi_{-t})_* Y_{\Phi_{tq}} \right\}_{t=0}.
\end{aligned} \tag{A.46}$$

i.e., we compare the vector $Y_{\Phi_{tq}}$ at that point with the result of pushing Y_q to the point of Φ_{tq} by means of the differential $\Phi_{t*} = D\Phi_t$ (Figs. A.8 and A.10).

It is proved that the commutator (the Lie bracket), $[X, Y]$, of two vector fields X and Y, with all of them $(X, Y, \mathscr{L}_X Y) \in \mathscr{V}_{\mathscr{F}}(M)$, satisfy

$$\mathscr{L}_X Y = [X, Y] = XY - YX. \tag{A.47}$$

A Lie derivative acting on $f \in \mathscr{F}(M)$ gives

$$\mathscr{L}_X Y_q(f) = [X, Y]_q(f) = [Y(f)]X_q - [X(f)]Y_q. \tag{A.48}$$

In local coordinates q^i the components of the Lie bracket are written

Fig. A.8 Geometrical interpretation of the Lie derivative of the vector field Y with respect to the vector field X which generates the flow Φ_{tq}

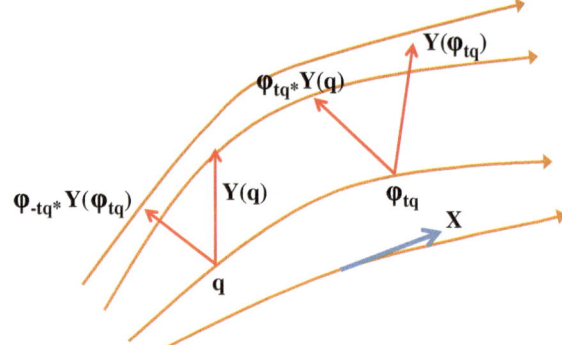

$$[X, Y]^i = \sum_j \left[\left(\frac{\partial Y^i}{\partial q^j} \right) X^j - \left(\frac{\partial X^i}{\partial q^j} \right) Y^j \right]. \qquad (A.49)$$

If we take $X^i = dq^i/dt = \dot{q}^i$ along the orbit, we can also write

$$[X, Y]^i = [\mathscr{L}_X Y]^i = \sum_j \left(\frac{\partial Y^i}{\partial q^j} \right) \dot{q}^j - \sum_j \left(\frac{\partial X^i}{\partial q^j} \right) Y^j$$

$$= \frac{dY^i}{dt} - \sum_j \left(\frac{\partial X^i}{\partial q^j} \right) Y^j. \qquad (A.50)$$

The vector field Y along an orbit of X is *invariant* if

$$Y_{\Phi_t q} = \Phi_{t*} Y_q. \qquad (A.51)$$

Then

Fig. A.9 Lie derivative of the vector field $Y = \delta q$ with respect to the vector field X which generates the flow Φ_{tq}

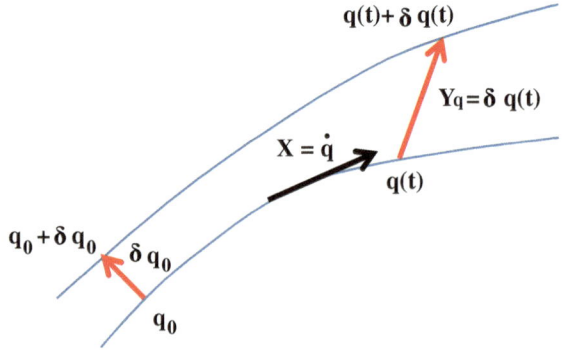

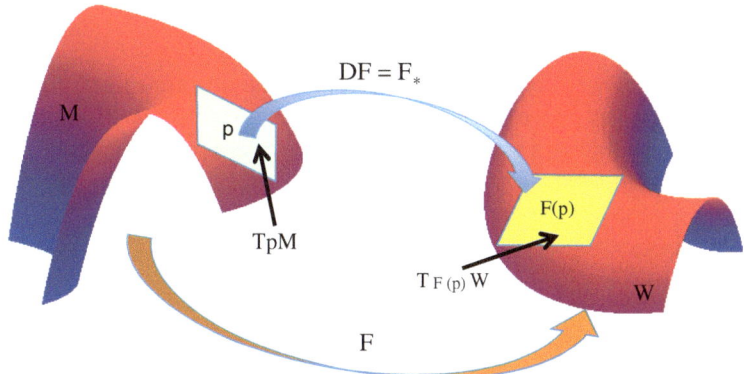

Fig. A.10 F is a smooth map between the manifolds M and W. $F_* = DF$ is a linear map of the tangent space $T_p M$ of M at the point p onto the tangent space $T_{F(p)} W$ at the point of $F(p)$

$$[\mathscr{L}_X Y]^i = \frac{dY^i}{dt} - \sum_j \left(\frac{\partial X^i}{\partial q^j}\right) Y^j = 0. \tag{A.52}$$

Even more, if we take a variation vector δq, an infinitesimally near point to q, to be the vector field Y (Fig. A.9), then

$$\frac{d(\delta q^i)}{dt} = X^i_{q+\delta q} - X^i_q$$
$$= \sum_j \left(\frac{\partial \dot{q}^i}{\partial q^j}\right) \delta q^j + h.o.t. \tag{A.53}$$

The higher order term (h.o.t.) is a function of the displacement δq at time t, which contains all the terms in the Taylor expansion larger than the first order. The derivatives are computed at the reference trajectory X_q. $Y = \delta q$ is called *Jacobi field* and the equations

$$\frac{dY^i}{dt} = \sum_j \left(\frac{\partial X^i}{\partial q^j}\right) Y^j, \tag{A.54}$$

variational equations.

In general, *the Lie bracket is preserved under diffeomorphisms.*

A.8 Cotangent (Dual) Space and Covectors

A linear functional on a vector space V is a scalar-valued function ω defined for every vector v, with the property

$$\omega(a_1 v_1 + a_2 v_2) = a_1 \omega(v_1) + a_2 \omega(v_2), \tag{A.55}$$

for real value numbers a_1, a_2 and vectors v_1, v_2. The set of all linear functionals acting on the tangent space at point p, $T_p M$, of the manifold M, forms the *dual space* of the manifold at the point p. This is the vector space of *covectors*, called *cotangent space* to M at the point p, and it is denoted as $T_p^* M$. The disjoint union of the cotangent spaces over all point p of M,

$$T^* M = \bigcup_{p \in M} T_p^* M, \tag{A.56}$$

is the *cotangent bundle* of M. Like the tangent bundle the cotangent bundle is a *differentiable manifold*.

The elements of $T_p^* M$ are denoted as ω_p and are named *differential forms of degree-1* or $1-$ *forms* and they are defined:

$$\omega : M \mapsto T^* M : p \mapsto \omega_p \in T_p^* M, \tag{A.57}$$

that assign to every point $p \in M$ an element $\omega_p \in T_p^* M$, which is a linear map of the tangent space $T_p M$ onto the real numbers, i.e., $\omega_p(v_p) \in \mathbb{R}$.

Taking a chart $(q(p) = (q^1(p), \ldots, q^n(p))$ an element of $T_p M$ has the representation $v = (v^1(q^1)\partial_1, \ldots, v^n(q^n)\partial_n)$ in the coordinate basis set ∂_i. The differential of coordinate q^i is dq^i and satisfies

$$dq^i(\partial_j\big|_p) = \frac{\partial_i}{\partial q^j}\big|_p dq^i = \delta^i_j. \tag{A.58}$$

Thus, we may conclude that the differentials dq^i can be considered as $1-$forms which consist the coordinate basis set for expanding any $1-$form

$$\omega = \sum_{i=1}^{n} \omega(\partial_i) dq^i. \tag{A.59}$$

The action of dq^i on a vector $v \in T_q M$ gives the component v^i of the vector;

$$dq^i(v) = dq^i \left(\sum_j v^j \partial_j \right) = \sum_j v^j dq^i(\partial_j) = \sum_j v^j \delta^i_j = v^i. \tag{A.60}$$

An example of $1-$form is the total differential of a function f

$$df : TM \to \mathbb{R} : df = \sum_{i=1}^{n} \frac{\partial f}{\partial q^i} dq^i = \sum_{i=1}^{n} \partial_i(f) dq^i. \tag{A.61}$$

Thus, the relation between tangent space and cotangent space of a manifold M at the point p is to map the vectors $v_p \in T_p M$ to real numbers via $1-$forms $\omega(v_p) \in T_p^* M$. The set $(dq^i |_p)$ is the coordinate basis for the cotangent vector space and the set $(\partial_i |_p)$ the coordinate basis for the tangent vector space $T_p M$.

A.8.1 Coordinate Transformation of 1−Forms

For a transformation of local coordinates, $w = F(q)$, the chain rule implies

$$dw^i = \sum_{j=1}^{n} \left(\frac{\partial F^i}{\partial q^j} \right) dq^j, \, i = 1, \ldots, n. \tag{A.62}$$

For the components of a general $1-$form the transformation is written

$$\sum_{i=1}^{n} \omega^w (\partial_i^w) dw^i = \sum_{i=1}^{n} \omega_i^w dw^i = \sum_{i=1}^{n} \omega_i^w \sum_{j=1}^{n} \left(\frac{\partial F^i}{\partial q^j} \right) dq^j. \tag{A.63}$$

Since,

$$\sum_{i=1}^{n} \omega_i^w dw^i = \sum_{j=1}^{n} \omega_j^q dq^j, \tag{A.64}$$

we conclude that

$$\omega_j^q = \sum_i \omega_i^w \left(\frac{\partial F^i}{\partial q^j} \right). \tag{A.65}$$

Taking the inverse of the Jacobian matrix we write the component transformation of a covector as

$$\omega_i^w = \sum_j \omega_j^q \left[\left(\frac{\partial F}{\partial q} \right)^{-1} \right]_i^j. \tag{A.66}$$

A.9 Exterior (Grassmann) Algebra

A.9.1 Exterior Product

A $k-$form is a *function* on a smooth manifold M

$$\overset{k}{\omega}: M \mapsto (T^*M)^k : p \mapsto \overset{k}{\omega}_p, \tag{A.67}$$

that assigns to each point $p \in M$ an element of $(T_p^*M)^k$, the $k-$fold *direct product* of the cotangent space. $\overset{k}{\omega}_p$ is the *exterior product* of k $1-$forms and it is a *multilinear, skew-symmetric* function from $(T_pM)^k$ onto the real numbers $\mathbb{R}$, i.e., it acts on k vector fields

$$\overset{k}{\omega}_p (v_1, \ldots, v_k) \in \mathbb{R}, \tag{A.68}$$

and is antisymmetric in all k arguments. A $k-$form using determinants and the symbol of *wedge* product ($\wedge$) is written

$$(\overset{1}{\omega}_1 \wedge \overset{1}{\omega}_2 \wedge \cdots \wedge \overset{1}{\omega}_k)(v_1, v_2, \ldots, v_k) = \frac{1}{k!} \begin{vmatrix} \overset{1}{\omega}_1 (v_1) & \cdots & \overset{1}{\omega}_1 (v_k) \\ \overset{1}{\omega}_2 (v_1) & \cdots & \overset{1}{\omega}_2 (v_k) \\ \cdot & \cdots & \cdot \\ \cdot & \cdots & \cdot \\ \cdot & \cdots & \cdot \\ \overset{1}{\omega}_{k-1} (v_1) & \cdots & \overset{1}{\omega}_{k-1} (v_k) \\ \overset{1}{\omega}_k (v_1) & \cdots & \overset{1}{\omega}_k (v_k) \end{vmatrix} \tag{A.69}$$

An (exterior) $k-$form is a covariant $k-$tensor, but not all covariant $k-$tensors are $k-$forms. The collection of all $k-$forms is a vector space.

Locally (in charts) any $k-$form can be written as a linear combination of the base forms,

$$\overset{k}{\omega} = \sum_{i_1 < i_2 < \cdots < i_k} \omega_{i_1 i_2 \ldots i_k} dq^{i_1} \wedge \cdots \wedge dq^{i_k}. \tag{A.70}$$

The coefficients are given by the action of $\overset{k}{\omega}$ onto the corresponding base vector fields

$$\omega_{i_1 i_2 \ldots i_k} = \overset{k}{\omega} (\partial_{i_1}, \ldots, \partial_{i_k}). \tag{A.71}$$

A.9.2 The Geometric Meaning of Forms in $\mathbb{R}^n$

For an Euclidean metric space, $\mathbb{R}^n$, with Cartesian coordinates, $(q^1, \ldots, q^n)$, the basis of the tangent space is $(\partial_1, \ldots, \partial_n)$ and that of the dual space the basis is $(dq^1, \ldots, dq^n)$. Then, for a pair of vectors (v, w) the action of the differentials is $dq^i(v) = v^i$, and $dq^j(w) = w^j$, i.e., the components of the vectors (v, w). It is valid

$$dq^i \wedge dq^j (v, w) = dq^i(v)dq^j(w) - dq^j(v)dq^i(w)$$
$$= v^i w^j - v^j w^i$$
$$= det \left| v^i w^j \right|. \tag{A.72}$$

This is $\pm$ the area of the parallelogram spanned by the projections of (v, w) vectors into (q^i, q^j) plane; the plus sign is used if the projections determine the same orientation of the plane as do ∂_i and ∂_j.

For a $k-$form

$$dq^{i_1} \wedge \cdots \wedge dq^{i_k}(v_1, \ldots, v_k), \tag{A.73}$$

is the $\pm k-$dimensional volume of the parallelepiped spanned by the projections of the v^i vectors into $(q^{i_1}, \ldots, q^{i_k})$ coordinate plane; the plus sign is used only if these projected vectors define the same orientation as do $(\partial_{i_1}, \ldots, \partial_{i_k})$.

A.9.3 Interior Product

If v is a vector and ω is a $k-$form, their *interior product* $(k - 1)-$form, $i_v \overset{k}{\omega}$, is defined by

$$i_v \overset{0}{\omega} = 0$$
$$i_v \overset{1}{\omega} = \omega(v) = \sum_j \omega_j v^j$$
$$i_v \overset{k}{\omega}(w_2, \ldots, w_k) = \overset{k}{\omega}(v, w_2, \ldots, w_k). \tag{A.74}$$

It is proved that

$$i_{v_1+v_2} = i_{v_1} + i_{v_2}$$
$$i_{av} = a i_v$$
$$i_v(\overset{k}{\alpha} \wedge \overset{l}{\beta}) = [i_v \overset{k}{\alpha}] \wedge \overset{l}{\beta} + (-1)^k \overset{k}{\alpha} \wedge [i_v \overset{l}{\beta}]. \tag{A.75}$$

A.9.4 Exterior Derivative

We consider smooth functions f on a manifold to be forms of degree zero. The *total differential*

$$df = \sum_{i=1}^{n} \frac{\partial f}{\partial q^i} dq^i, \tag{A.76}$$

is an $1-$form expanded in the basis forms dq^i and coefficients the partial derivatives of f, i.e., the action of $df \equiv \overset{1}{\omega}$ onto the corresponding base vector fields, $(\partial_{i_1}, \ldots, \partial_{i_n})$.

Generally, the exterior or Cartan derivative of an exterior product of two forms is defined as

$$d(\overset{k}{\omega} \wedge \overset{l}{\omega}) = (d \overset{k}{\omega}) \wedge \overset{l}{\omega} + (-1)^k \overset{k}{\omega} \wedge (d \overset{l}{\omega}), \tag{A.77}$$

where

$$d \overset{k}{\omega} = \sum_{i_1 < \cdots < i_k} d\omega_{i_1 \ldots i_k}(q^1, \ldots, q^n) \wedge dq^{i_1} \wedge \cdots \wedge dq^{i_k}. \tag{A.78}$$

$d\omega_{i_1 \ldots i_k}(q^1, \ldots, q^n)$ is the total differential of the component $\omega_{i_1 \ldots i_k}(q^1, \ldots, q^n)$ function. The d operation is said to be an *antiderivation*.

In other words, the exterior derivative maps smooth $k-$forms onto $(k+1)-$forms

$$d : \overset{k}{\omega} \to \overset{k+1}{\omega}. \tag{A.79}$$

It is proved that

$$d \circ d = 0. \tag{A.80}$$

A.9.5 Pull-Back Differential Forms and Push-Forward Vector Fields

How are vectors from a tangent space of a manifold transformed to another manifold connected by a differentiable map (or diffeomorphism)? Similarly, how are forms transformed between cotangent spaces of two manifolds connected by a differentiable map? To perform such operations we *push-forward* vectors and *pull-back* forms.

Let

$$F : M^m \mapsto W^r, \tag{A.81}$$

is a differentiable map between the two manifolds M of dimension m and W of dimension r, then the *push-forward* of a vector field $V_p \in T_p M$ on the manifold M to the vector field $V'_{F(p)} \in T_{F(p)} W$, the differential of F, is denoted as $DF = F_*$ and its action is

$$V'_{F(p)} = (F_*v)_{F(p)} = V_p \circ F = F_*(V_p). \tag{A.82}$$

If $f : W \to \mathbb{R}$ then

$$V'_{F(p)} = F_* V_p$$
$$V'_{F(p)}(f) = V_p(f \circ F). \tag{A.83}$$

In local coordinates the transformation equations for the vector components are as those of a coordinate transformation $s^i = F^i(q)$, where q are the coordinates for the manifold M and s the coordinates for manifold W.

$$V^i = \sum_j V'^j \frac{\partial q^i}{\partial s^j} \tag{A.84}$$

$$V'^j = \sum_i V^i \frac{\partial s^j}{\partial q^i} = \sum_i V^i \frac{\partial F^j}{\partial q^i}. \tag{A.85}$$

The pull-back operation is defined as

$$[F^* \overset{k}{\omega}_p](v_1, \ldots, v_k) = \overset{k}{\omega}_{F(p)} (F_* v_1, \ldots, F_* v_k), \tag{A.86}$$

where $(v_1, \ldots, v_k)$ are tangent vectors to M and $p \in M$. In other words

$$M \overset{F}{\to} W \overset{\omega}{\to} T^*W \overset{(F_*)^*}{\to} T^*M. \tag{A.87}$$

Properties of the pull-back operation

$$F^*(\overset{k}{\omega} \wedge \overset{l}{\omega}) = (F^* \overset{k}{\omega}) \wedge (F^* \overset{l}{\omega}), \tag{A.88}$$

$$F^*(d \overset{k}{\omega}) = d(F^* \overset{k}{\omega}). \tag{A.89}$$

If we define a chart $q(p) = (q^1(p), \ldots, q^m(p))^T$ at $p \in M$, a chart $s(F(p)) = (s^1, \ldots, s^r)^T$ at $F(p) \in W$, and the components of function $F = (F^1, \ldots, F^m)^T$ with $s^i = F^i(q^1, \ldots, q^m)$, the coordinate representation of the pull-back of the k-form, $\overset{k}{\omega}$, of the manifold W is

$$\overset{k}{\omega} = \sum_{i_1 < i_2 < \cdots < i_k} \omega_{i_1 i_2 \ldots i_k} ds^{i_1} \wedge \cdots \wedge ds^{i_k}. \tag{A.90}$$

The pull-back of $\overset{k}{\omega}$ is then

$$F^* \overset{k}{\omega} = \sum_{i_1 < i_2 < \cdots < i_k} (\omega_{i_1 i_2 \ldots i_k} \circ F) dF^{i_1} \wedge \cdots \wedge dF^{i_k}. \tag{A.91}$$

Transforming dF^i to the coordinates q with Jacobians results in

$$F^* \overset{k}{\omega} = \sum_{i_1 < i_2 < \cdots < i_k} \sum_{j_1 < j_2 < \cdots < j_k} (\omega_{i_1 i_2 \ldots i_k} \circ F) \frac{\partial(F^{i_1}, \ldots, F^{i_k})}{\partial(q^{j_1}, \ldots, q^{j_k})} dq^{j_1} \wedge \cdots \wedge dq^{j_k}. \tag{A.92}$$

A.9.5.1 Example 0−Form

Let us see how the above formulation applies to functions (0−forms). $F : M \to W$ be a differentiable map. In local coordinates q for M and s for W we have $s^i = F^i(q)$, or briefly $s = s(q)$. If $f : W \to \mathbb{R}$ is a smooth function on W we define its pull-back to M, written $F^* f$, to be the *composition* $(f \circ F)(q) \to \mathbb{R}$, that is, $M \overset{F}{\to} W \overset{f}{\to} \mathbb{R}$.

$$(F^* f)(q) = (f \circ F)(q) = f[s(q)]. \tag{A.93}$$

A vector $v \in T_q M$ of M acts on the pull-back of a function in the following way

$$v(F^* f) = v\{f[s(q)]\} = \sum_i v^i \partial_i \{f[s(q)]\}$$

$$= \sum_i v^i \sum_j \left(\frac{\partial s^j}{\partial q^i}\right) \left(\frac{\partial f}{\partial s^j}\right), \tag{A.94}$$

or

$$v(F^* f) = (F_* v)(f) = df(F_* v). \tag{A.95}$$

A.9.6 Lie Derivative of a Form

$\overset{k}{\omega}$ is a k−form, $\Phi_t(q)$ a local 1-parameter group of diffeomorphisms (flow) with X the associated vector field ($\dot{\Phi}(q) = X_q$). Then, Φ_{tq} is the point at time t along the *integral curve* of X, the *orbit* of q, that starts at time $t = 0$ at the point q. Putting $\overset{k}{\omega}(q) = \omega_q$, the Lie derivative is defined by employing the pull-back of the flow, Φ_t^*,

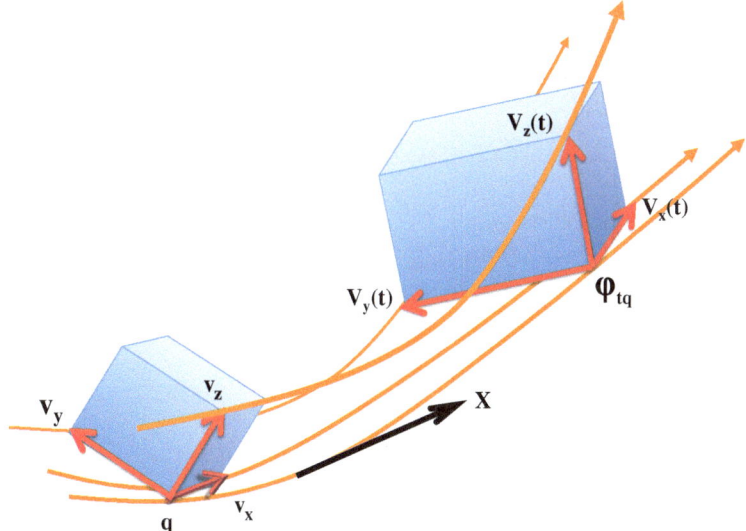

Fig. A.11 Interpretation of the Lie derivative of a form as the rate of deformation of the volume of a parallelepiped along the flow Φ_t

$$\mathscr{L}_X \overset{k}{\omega} = \frac{d}{dt}\left[\Phi_t^* \overset{k}{\omega}\right]_{t=0}$$

$$= \lim_{t\to 0}\frac{\Phi_t^*\omega_{\Phi_t(q)} - \omega_q}{t}. \qquad (A.96)$$

Let $v_1, \ldots, v_k$ be vector fields at q, then

$$\left[\frac{d}{dt}\Phi_t^* \overset{k}{\omega}\right](v_1, \ldots, v_k) = \frac{d}{dt}\left[\Phi_t^* \overset{k}{\omega}(v_1, \ldots, v_k)\right]$$

$$= \frac{d}{dt}\left\{\overset{k}{\omega}[\Phi_{t*}v_1, \ldots, \Phi_{t*}v_k]\right\}. \qquad (A.97)$$

For *invariant vector fields* along the orbit through q, $\Phi_{t*}v_q = v_{\Phi_t q}$, then we can write

$$\mathscr{L}_X \overset{k}{\omega} = \left\{\frac{d}{dt}\left[\overset{k}{\omega}_{\Phi_t(q)}(v_1, \ldots, v_k)\right]\right\}_{t=0}. \qquad (A.98)$$

Thus, If we interpret forms as volumes of parallelepipeds spanned by the vectors $(v_1, \ldots, v_k)$ (Sect. A.9.2), then the Lie derivative $\mathscr{L}_X \overset{k}{\omega}$ measures the deformation of this volume along the flow $\Phi_t(q)$, Fig. A.11.

A.9.6.1 Example : The Lie Derivative of a Function (0−Form)

We have seen that if X is the vector field that generates the local flow $\Phi_t(q)$, then

$$\mathscr{L}_X f = X(f) = \sum_i X^i \partial_i f. \tag{A.99}$$

$$\mathscr{L}_X \overset{0}{\omega} = \mathscr{L}_X f = \frac{d}{dt} f\,[\Phi_t(q)]_{t=0} = \frac{d}{dt}\,[\Phi_t^* f]_{t=0} = \sum_i \dot{q}^i \frac{\partial f}{\partial q^i}. \tag{A.100}$$

A.9.6.2 Properties of the Lie Derivatives for Forms

(i) $\mathscr{L}_X(\overset{k}{\omega} \wedge \overset{l}{\omega}) = (\mathscr{L}_X \overset{k}{\omega}) \wedge \overset{l}{\omega} + \overset{k}{\omega} \wedge (\mathscr{L}_X \overset{l}{\omega})$
(ii) $\mathscr{L}_X \circ d = d \circ \mathscr{L}_X$

A.9.7 Closed Forms and Exact Forms

A form $\overset{k}{\omega}$ is *closed* if

$$d\,\overset{k}{\omega} = 0. \tag{A.101}$$

We can prove

1. $d(\overset{0}{\omega}) = 0 \iff \overset{0}{\omega}$ is a constant function.
2. $d(\overset{1}{\omega}) = 0 \iff (\partial_i \omega_j - \partial_j \omega_i) = 0.$
3. $d(\overset{2}{\omega}) = 0 \iff (\partial_i \omega_{jk} + \partial_j \omega_{ki} + \partial_k \omega_{ij}) = 0.$

A form $\overset{k}{\omega}$ is *exact* if

$$\overset{k}{\omega} = d\,\overset{k-1}{\omega}. \tag{A.102}$$

It is also valid

1. Every exact form is closed.
2. The product of two closed forms is closed.
3. The product of a closed form and an exact form is exact.
4. The integral of an exact form over an oriented closed manifold (i.e., compact without boundary) is 0.
5. The integral of a closed form over the boundary of an oriented compact manifold is 0.

A.9.8 Stoke's Theorem

We may consider forms as densities, and thus, they can be the natural integrands in physics.

A $m-$*manifold with boundary M* has an interior that is a genuine $m-$manifold, and a boundary usually written, ∂M.

Theorem A.4 (Stokes's Theorem [10])

Let $V \subset M$ be a compact oriented manifold with boundary ∂V. Let $\overset{k-1}{\omega}$ be a continuously differentiable $(k-1)-$form on M. Then

$$\int\limits_V d\,\overset{k-1}{\omega} = \int\limits_{\partial V} \overset{k-1}{\omega}\,. \tag{A.103}$$

A.9.9 Poincaré Lemma

Theorem A.5 (Poincaré Lemma)

If $d(\overset{k}{\omega}) = 0$, $k \geq 1$, in a neighborhood U of $q \in M$, then there is perhaps a smaller neighborhood U' of q and a $(k-1)-$form, $\overset{k-1}{\omega}$, such that

$$\overset{k}{\omega} = d(\overset{k-1}{\omega}). \tag{A.104}$$

A.10 Hamiltonian Normal Forms Around Equilibria

Normal form expansion of Hamiltonians is a powerful method for constructing phase space structures such as tori, NHIM and stable and unstable manifolds. The aim of normal form approximation is to build a series of symplectic variable transformations, which successively and up to a predefined order transform the Hamiltonian to its 'normal form', a property given below. The Taylor expansion is made around equilibria and periodic orbits and the methodology is described in several books of nonlinear mechanics. Here, we briefly present the theory for a stable equilibrium as has been described by Meyer [7] (see also [1]).

First, we assume that we can expand the global Hamiltonian in a Taylor series up to the required order and around the equilibrium x_0, which is taken to be stable and the origin of the coordinate system. Then, assuming the constant term to be zero the Taylor series is written as

$$H(x, \varepsilon) = \sum_{l=0}^{\infty} \frac{\varepsilon^l}{l!} H_l^0(x). \tag{A.105}$$

H_l^0 is a homogeneous polynomial of degree $l + 2$ in the variables $x = (q^1, \ldots, q^n,$ $p_1, \ldots, p_n)^T$, and ε is a scaling factor (dilation), related to the magnitude of the neighbourhood that we take around the equilibrium.

$H_0^0(x)$ denotes the quadratic part of the Hamiltonian, $H_0^0(x) = \frac{1}{2} x^T [\partial^2 H] x$, where $[\partial^2 H]$ is the Hessian of the Hamiltonian evaluated at the equilibrium point. The linearized Hamiltonian vector field is defined by

$$\dot{x}(t) = J[\partial^2 H] x(t) = A x(t). \tag{A.106}$$

By diagonalizing the matrix, $A = J[\partial^2 H]$, we obtain the *normal coordinates* of the molecule which render the quadratic Hamiltonian into a harmonic one [4]

$$H_0^0 = \frac{1}{2} \sum_{i=1}^{n} (p_i^2 + \omega_i^2 q^{i2}), \tag{A.107}$$

where ω_i are the eigenfrequencies of the normal modes.

We can scale the normal coordinates by the symplectic transformation

$$Q^i = \sqrt{\omega_i} q^i \tag{A.108}$$

$$P_i = p_i / \sqrt{\omega_i}, \quad i = 1, \ldots, n. \tag{A.109}$$

Then, the Hamiltonian H_0^0 takes the form

$$H_0^0 = \frac{1}{2} \sum_{i=1}^{n} \omega_i (P_i^2 + Q^{i2}). \tag{A.110}$$

We further transform to complex variables by the normalized symplectic transformation

$$z^i = \frac{1}{\sqrt{2}} (Q^i - {}_1 P_i)$$

$$w_i = \frac{1}{\sqrt{2}} (-{}_1 Q^i + P_i), \quad i = 1, \ldots, n, \tag{A.111}$$

where ${}_1 = \sqrt{-1}$. Then, the wedge products in the Hamiltonian symplectic 2−form are transformed as (Eq. 2.57)

$$dz^i \wedge dw_i = dQ^i \wedge dP_i. \tag{A.112}$$

The quadratic Hamiltonian becomes

---

[4] To avoid using many symbols we use (q, p) to ascribe both the internal and the normal coordinates.

$$H_0^0 = \sum_{i=1}^{n} \imath \omega_i \, z^i \, w_i. \tag{A.113}$$

We introduce the action-angle variables (I_i, ϕ^i)

$$z^i = \sqrt{I_i} \exp(\imath \phi^i)$$
$$w_i = -\imath \sqrt{I_i} \exp(-\imath \phi^i) \quad \phi \in [0, 2\pi], \quad I_i \in (0, \infty), \tag{A.114}$$

and the Hamiltonian in action-angle variables is transformed to

$$H(I, \phi, \varepsilon) = \sum_{l=0}^{\infty} \frac{\varepsilon^l}{l!} H_l^0(I, \phi). \tag{A.115}$$

The quadratic part is

$$H_0^0 = \sum_{i=1}^{n} \omega_i I_i, \tag{A.116}$$

and the other terms are written into the form

$$\sum_{\|m\| \le l+2} h_m^l(I) e^{\imath <m, \phi>}, \tag{A.117}$$

where $< m, \phi > = \sum_{i=1}^{n} m_i \phi^i$, $m_i \in \mathbb{Z}$ and $\| m \| = \sum_{i=1}^{n} |m_i|$. h_m^l are homogeneous polynomials of degree $(1 + l/2)$ in I_i.

For incommensurable harmonic frequencies we can obtain integrable Hamiltonians which incorporate higher order terms of Eq. A.115 by employing normal form transformations. It is proved that a symplectic transformation of the variable x, where now x signifies the complex coordinates $x = (z^1, \ldots, z^n, w_1, \ldots, w_n)^T$

$$x = \mathcal{X}(y), \tag{A.118}$$

with inverse

$$y = \mathcal{Y}(x), \tag{A.119}$$

leads to the normal form Hamiltonian $H^\ominus(y) = H[\mathcal{X}(y)]$, such as

$$H^\ominus(y) = \sum_{l=0}^{\infty} \frac{\varepsilon^l}{l!} H_0^l(y). \tag{A.120}$$

$H_0^l(y)$ is a homogeneous polynomial of degree $l + 2$ in the normal form variables y. The vector y collectively denotes the normal form coordinates F^i and their conjugate momenta P_{F^i} in a complex conjugate manner analogously to z and w, $y = [(F^1 - {}_1P_{F^1})/\sqrt{2}, \ldots, (F^n - {}_1P_{F^n})/\sqrt{2}, (-{}_1F^1 + P_{F^1})/\sqrt{2}, \ldots, (-{}_1F^n + P_{F^n})/\sqrt{2}]$ (see also Sect. 4.1.2).

Taking $H_0^0(y) = H_0^0(x)$, the higher order terms of H_0^l satisfy

$$\{H_0^l, H_0^0\} = 0, \quad l = 0, 1, 2, \ldots, L, \tag{A.121}$$

and the Hamiltonian $H^\ominus$ is in normal form through degree $L+2$. The Poisson bracket above is defined as

$$\{H_0^l, H_0^0\} = \left(\frac{\partial H_0^l}{\partial y}\right)^T J \left(\frac{\partial H_0^0}{\partial y}\right)$$

$$\equiv \sum_{\mu,\nu=1}^{2n} \left(\partial_\mu H_0^l\right) J^{\mu\nu} \left(\partial_\nu H_0^0\right). \tag{A.122}$$

J is the symplectic matrix (Eq. 2.49). Eq. A.121 implies that $H_0^l(y)$ are conserved quantities along the trajectories of the quadratic Hamiltonian H_0^0. The following theorem has been proved by Meyer [6, 7].

Theorem A.6 *Let*

$$H(x, \varepsilon) = \sum_{l=0}^{\infty} \frac{\varepsilon^l}{l!} H_i^0(x). \tag{A.123}$$

be a formal Hamiltonian where $H_i^0(x)$ is a homogeneous polynomial of degree $(l+2)$ and $H_0^0(x) = \frac{1}{2}x^T \partial^2 H x$ the quadratic Hamiltonian with the matrix $A = J\partial^2 H(x)$ to have incommensurable eigenvalues. Then, there exists a real formal canonical transformation $x = \mathscr{X}(y, \varepsilon)$ which transforms H to $H^\ominus$, where

$$H^\ominus(y) = \sum_{l=0}^{\infty} \frac{\varepsilon^l}{l!} H_0^l(y). \tag{A.124}$$

where H_0^l is a homogeneous polynonial of degree $(l + 2)$ and

$$H_0^l(e^{At} y) = H_0^l(y), \tag{A.125}$$

for all t and y.

In practice, to find the normal form coordinates we take the following steps. We are seeking for *near identity* coordinate transformations

$$x = y + \sum_{l=1}^{\infty} \frac{\varepsilon^l}{l!} y_0^l. \tag{A.126}$$

This is obtained by solving Hamilton's equation to find the generating function W

$$\frac{dx}{d\varepsilon} = J\partial W(\varepsilon, x), \tag{A.127}$$

and initial conditions $x(0) = y$. W is also expanded as

$$W(x, \varepsilon) = \sum_{l=0}^{\infty} \frac{\varepsilon^l}{l!} W_{l+1}(x). \tag{A.128}$$

$W_l(x)$ is a homogeneous polynomial of degree $(l + 2)$. The coordinates y_0^l are then calculated by the recursive formula

$$y_i^{(j)} = y_{i+1}^{(j-1)} + \sum_{k=0}^{i} \binom{i}{k} \{y_k^{(j-1)}, W_{i+1-k}\}, \tag{A.129}$$

where $i \geq 0$, $j \geq 1$, $y_i^{(0)} \equiv 0$ for $i \geq 1$ and $y_0^{(0)} \equiv y$.

To construct the normal forms for $l > 0$ we compute the Lie series via the Lie-Deprit algorithm [2]. The method introduces the functions H_i^j, which are homogeneous polynomials of degree $i + j + 2$ and with $i, j \geq 0$, and are obtained by the recurrence equation

$$H_i^j = H_{i+1}^{j-1} + \sum_{k=0}^{i} \binom{i}{k} \{H_{i-k}^{j-1}, W_{k+1}\}. \tag{A.130}$$

$\{H_{i-k}^{j-1}, W_{k+1}\}$ are Poisson brackets and the binomial coefficients are written

$$\binom{i}{k} = \frac{i!}{k!(i-k)!}. \tag{A.131}$$

The functions W_i are found by solving the *homological equation*

$$\hat{H}_i + \{H_0^0, W_i\} = 0, \tag{A.132}$$

where $\hat{H}_i$ collects all terms which are not in normal forms. For details see [1, 7].

Similar formulae can also be extracted to calculate the inverse transformation

$$y = x + \sum_{l=1}^{\infty} \frac{\varepsilon^l}{l!} x_0^l, \tag{A.133}$$

and

$$x_i^{(j)} = x_{i-1}^{(j+1)} - \sum_{k=0}^{i-1} \binom{i-1}{k} \{x_{i-k-1}^{(j)}, W_{k+1}\}, \qquad \text{(A.134)}$$

with $i \geq 1$, $j \geq 0$, $x_{(0)}^i \equiv 0$ for $i \geq 1$ and $x_0^{(0)} \equiv x$.

Having obtained the normal form coordinates at a pre-specified order of accuracy, the constants of motion (action variables) are calculated by

$$I_i = (F^{i2} + P_{Fi}^2)/2. \qquad \text{(A.135)}$$

The normal form method can be applied to unstable equilibria as well, for example at a saddle [1], and then, the actions of the unstable degrees of freedom are expressed by the equation

$$J_j = F^j P_{Fj}. \qquad \text{(A.136)}$$

References

1. Collins P, Burbanks A, Wiggins S, Waalkens H, Schubert R (2008) Background and documentation of software for computing Hamiltonian normal forms, 1st edn. School of Mathematics, University of Bristol, University Walk, Bristol BS8 1TW
2. Deprit A (1969) Canonical transformations depending on a small parameter. Cel Mech 1:12–30
3. Frankel T (2004) The geometry of physics: an introduction. Cambridge University Press, The Edinburgh Building, Cambridge, UK (CB2 2RU)
4. Ginoux JM (2009) Differential geometry applied to dynamical systems. Nonlinear Science, World Scientific Publishing Co., Pte. Ltd., 5 Toh Tuck Link, Singapore, 596224
5. Halmos PR (1950) Measure theory. International Students Editions, Van Nostrand Reinhold Company, INC., New York, USA (450 West 33rd Street, 10001)
6. Meyer KR (1974) Normal forms for Hamiltonian systems. Celest Mech 9:517–522
7. Meyer KR, Hall GR, Offin D (2009) Introduction to Hamiltonian dynamical systems and the n-body problem, applied mathematical sciences, vol 90, 2nd edn. Springer, Heidelberg
8. Nakahara M (2003) Geometry, topology and physics. Graduate student series in physics, Taylor and Francis. 2 Park Square, Milton Park, Abingdon
9. Poryteous IR (1969) Topological geometry. The new university mathematics series. Van Nostrand Reinhold Company, INC., London (Windsor House, 46 Victoria Street, S.W.1, Great Britain)
10. Spivak M (1965) Calculus on manifolds: a modern approach to classical theorems of advanced calculus. Addison-Wisley, Massachusetts

Index

© The Author(s) 2014
S.C. Farantos, *Nonlinear Hamiltonian Mechanics Applied to Molecular Dynamics*,
SpringerBriefs in Electrical and Magnetic Properties of Atoms, Molecules, and Clusters,
DOI 10.1007/978-3-319-09988-0